Critical Perspectives on the DSM 5

2015

Critical Perspectives on the DSM 5

Compiled by Nora L. Ishibashi

2015

First Printing: 2015

ISBN 978-1-329-20132-3

Nora L. Ishibashi, Ph.D., P.C.
405 North Wabash Avenue, Suite 1303
Chicago, IL 60611

www.noraishibashi com

Ordering information:
This book can be ordered from http://www.lulu.com

Table of Contents

Introduction

Nora L. Ishibashi, Ph.D.

This book compiles papers written by both first-year master's degree students and undergraduate students in the Social Work Department at Loyola University of Chicago. In the Human Behavior in the Social Environment class, we study types of psychological distress, ways of assessing dysfunction, and implications of different viewpoints. Students apply theoretical and social viewpoints to our ways of responding to distress in our clients.

The *Diagnostic and Statistical Manual of Mental Disorders, Fifth Edition*, is the catalog of symptoms and diagnoses most often used by mental health professionals in the United States. It is revised and reissued as social changes alter our views of what constitutes normal behavior and what constitutes dysfunctional behavior. Although it has been used for many years and continues to be the standard for mental health diagnosis, there is much controversy surrounding its definitions and categories.

In these papers, students address some of the problems inherent in a specific diagnostic category from the viewpoint of different theoretical perspectives. The analyses are designed to expand our understanding of the unique individual stories that unfold in our relationships with clients. Social work is a profession with one hundred years of experience assessing the complex lives of individuals and groups and refining methods of joining with those clients in constructive ways. As part of our mission, we continue to critically examine our knowledge base and our methods of action.

We live in the social times we are in. We express the norms and values of our communities and nations. When there is a lack of fit between a person or group and the social environment in which they find themselves, social workers bring understanding, problem solving, and caregiving skills to bear on the situation. The evolution of the DSM over time reflects social changes and reminds us that these definitions of dysfunction are fluid, specific to the historic moment, and reflective of our cultural assumptions.

In our discussions we have considered the benefits and disadvantages of using a "diagnosis" to describe a person's distress. We have looked at arguments that criticize diagnosis as reductionistic and at arguments that recognize the functional benefits to the client of having a diagnosis.

Here you will find the reflective thoughtfulness and the sincere desire to be helpful of a new generation of social workers. These papers represent powerful expressions of students on the brink of entering the social world in order to advocate for the vulnerable, empower the troubled, and encourage the dispirited.

Sign Here: Enlisting to Receive Posttraumatic Stress Disorder

Allison T. Chan

Introduction

In today's Veteran Administration (VA) Healthcare system, the diagnosis of Posttraumatic stress disorder (PTSD) has become synonymous with mental illness. In the January 2015 publication of health statistics for veterans' access to care at VA health care centers, 351,422 veterans of the Operation Iraqi Freedom (OIF), Operation Enduring Freedom (OEF) and Operation New Dawn (OND) conflicts had received a diagnosis of PTSD. (DOD 2015). This alarming number accounts for over 31% of the overall population of OIF/OEF/OND veterans to seek health care at a VA facility in the past 12 years since the post 9-11 conflicts began. With close to two million military service members participating in at least one deployment in the past 12 years, the numbers of PTSD diagnoses will continue to rise as more of the veterans seek medical assistance in the years to come.

In the 2013 publication of the updated version of the American Psychiatric Association's *Diagnostic and Statistical Manual of Mental Disorders* Fifth Edition (DSM-5), the prevalence rate for the trauma based disorder was estimated to be at 8.7% for lifetime risk in the United States (APA 2013). While the DSM-5 does acknowledge the rates for those exposed to combat experiences are significantly higher than other trauma experiences, almost one in three veterans seen for health care is being diagnosed with PTSD. Do the authors and contributors to the diagnositic criteria need to re-evaluate their subjective criteria? With the rates steadily rising and the possibility of the undiagnosed hundreds of thousands veterans who have not yet sought health care out there, could the symptomatic state of those with PTSD be the norm for veterans? Should the widely recognized sub-culture of the military be exempt from such a diagnosis?

ff er

3

Defining the Disorder

It is estimated that over 89% of the population will experience at least one traumatic event during their lifespan (Breslau 2009). The authors of the DSM-5 decided to distinguish trauma and stressor-related disorders into their own chapter (Gabbard 2014). Previous editions included diagnoses such as PTSD in the cluster of Anxiety Disorders (APA 2000). More attention is being given to this group of diagnoses that is unlike any other diagnosis in that it must be preceded by an event, the exposure to trauma.

Criterion A in DSM-5 defines trauma as the "exposure to actual or threatened death, serious injury, or sexual violence in one (or more) of the following ways: directly experiencing the traumatic event(s), witnessing, in person, the event(s) as it occurred to others, learning that the traumatic event(s) occurred to close family member or close friend or experiencing repeated or extreme exposure to adverse details of the traumatic event(s)"(APA 2013, p. 271). The previous edition of the DSM included a two-fold criterion for the definition of trauma that added the requirement for an intense emotional response ("fear, helplessness, or horror", APA, 2000, p. 467). Friedman (2013) explains that one of the reasons criterion A_2 was removed from the current edition of the DSM is the phenomenal response of the absence of the intense emotional response to trauma from professionals such as military, police and firemen. Professionals such as the military report their past training teaches them to instinctually react in an unemotional manner during traumatic events (Friedman, Resick, Bryant, & Brewin, 2011).

The symptoms of the disorder are distinguished into four categories: intrusion, avoidance, negative alterations and arousal and reactivity alterations (APA 2013). Intrusion symptoms commonly entail re-experiencing the trauma; this could include nightmares, recalling distressing memories, and flashbacks. Typically these symptoms may appear after being triggered by familiar situations or senses from the trauma, like smells, tastes, or sights. Intrinsic memories may also appear on anniversaries of the event. Avoidance symptoms are expressed by the person avoiding places, people, or situations that can remind them of their trauma. A man who is mugged at gunpoint on the corner of Chicago and Franklin Avenues may do everything in his power to avoid walking by that same corner even if he has to take a longer route to get to work. Exhibiting negative beliefs such as "I am not safe anywhere" or constant states of fear, guilt, horror, anger or shame are some of the common symptoms classified in the negative mood alteration category. Lastly the category of arousal or reactivity is conveyed through startled responses, angry out-outburst or sleep disturbances.

Criterion G in the PTSD diagnosis states "The disturbance causes clinically significant distress or impairment in social, occupational, or other important areas of functioning" (APA 2013, p. 272). This criterion is meant to assist in the distinction of the symptoms experienced after a trauma from a normal reaction to an abnormal or pathological reaction (APA 2013). Diagnosticians must take into account the military as its own culture and society with its own set of values, norms and rules before deciding what impact the symptoms described above have on the individual's daily functioning.

Symbolic Interaction Theory in the Military

Symbolic Interaction is a theory that is attributed primarily to sociologist George Herbert Mead in the early 1900s (Blumer 1986). Blumer describes the three main premises for the theory: people act towards other people, places and objects based on the meaning they have for them, the meaning the person has for their surroundings comes out of their social interactions with their surroundings, and the meaning the person learns is then developed into their own interpretation that they go on to use in dealing with their everyday environment (Blumer 1986). Many sociological theories share the first premise that people act based on their understanding of the meaning they have for their surroundings, but symbolic interaction is unique in its second premise, that the meaning people give to the objects within their environment is based off of the meaning they have learned from their environment (Blumer 1986).

William C. Cockerham writes about the military and symbolic interaction in Chapter 20 of *Handbook of Symbolic Interactionism*. He gives life to the framework of the theory as it applies to the unique subculture of the military stating, "The mission of the military is to destroy the enemy" (Reynolds & Herman-Kinney, p. 491). The military of a country becomes a powerful force when it is united in its mission and goals to accomplish that mission. Military members are trained through a specific set of guidelines and values, aimed at making them successful. These guidelines and values become the shared lens with which they come to view their surroundings and establish an understanding of their interactions within their society and culture. Members come to construct a set of worldviews unique to their culture that are necessary for survival in their day-to-day environment (Reynolds & Herman-Kinney 2003). As training becomes indoctrinated in their minds, they no longer distinguish between their old and new societies. The mentality commonly dictated by commanders is to train as we fight and fight as we train.

The theory of symbolic interaction in the military has only a minute amount of research collected by sociologists. Cockerham attributes the lack of research to an absence of researchers who have gone into the military, specifically to study the society (Reynolds & Herman-Kinney 2003). For research to be accurately collected the expert must either join the military to study the interactions and write about their experience after their service is concluded or gain approval from appropriate channels to go incognito into a unit, neither of which has been completed in large numbers. I, myself, having been a member of the military for 12 years find my first-hand experience insightful and appropriate to interject into the context of this reflection.

On the battlefield, the survival of a military member and their unit is accomplished by adhering to the training and beliefs we have come to know as instinctual. The United States Army bases its core beliefs on four actions that we refer to as the "Warrior Ethos:" I will always place the mission first, I will never accept defeat, I will never quit, and I will never leave a fallen comrade. When the creeds and ethos are no longer just recited by loud thunderous voices at a morning formation of troops but instead become entrenched in our beings through reinforcement of combat exposure, the understanding of symbolic interaction in the military's culture comes to light.

Through rigorous training such as firing weapons, hand to hand combat, and learning attack and evasion skills, the military begins to assign new meanings and understandings to our day-to-day life based on the culture's understanding of the people, places and objects in our environment. To prepare for war, we must believe what psychologists may define as "faulty core beliefs," such as "the world is a dangerous place" and "the people around us are out to get us." In order to survive, we have to be constantly vigilant to every sensation around us. We are trained to live in a constant state of fight or flight. We have assigned new meanings to the objects and events in our lives, meanings that are necessary to accomplish the mission.

Discussion

Using the premises of Symbolic Interaction Theory while looking at the theory with respect to the military as its own subculture, I must question the framework of the DSM-5's diagnostic criteria when it leads to over 30%, as a modest estimate, of veterans from the past 12 years receiving a diagnosis of PTSD. These symptoms, or learned behaviors, taken into context of the United States society as a whole, could be perceived as negative or an impairment to social functioning; understanding the presenting symptoms within the context of the military culture and the symbolic meaning the symptoms are assigned may prove to be beneficial rather than detrimental to the individual's functioning. I argue that each of these symptoms serves a purpose in the survivability of the military member and although may appear exhausted by the traumatic exposure of combat, in fact are learned responses that leaders have armed their soldiers with to prevent loss of life and promote mission accomplishment.

Many people can understand how the commonly attributed symptoms of PTSD negatively impact a person, but let us look at the positive impact they have for someone in the military. The first category is intrusion symptoms, include re-experiencing thoughts such as flashbacks. To respond to a horrific event such as being in a firefight with the enemy, a member of the military must act on the instinct they have learned. If they have been in a similar situation before they are able to recall details and make necessary plans to how they would proceed differently or similarly to protect themselves and others. Captain Smith may have a flashback every month of the combat patrol he lead in Iraq that lead to the loss of one of his soldiers. Although at times difficult to emotionally remember, Captain Smith needs to analyze his decisions that day as a leader and provide himself feedback of what he may do differently next time. He needs to remember how the experience played out; in the future the replaying of events could help to save a life.

Avoiding places that remind a trauma survivor of his or her experience may be beneficial to prevent the soldier from becoming emotionally overstimulated by the past event. A soldier needs to retain a certain level of dissociation from their emotions to be effective on the battlefield. If the soldier thinks that going to see a certain movie with her friends is going to become too overwhelming emotionally, it is not a bad decision for her to avoid the movie and find another activity. A soldier's job requires him or her to remain calm and collected; he or she should avoid any stimulus that will upset his or her frame of mind. If that same soldier has to go to a weapons range the next day, it is not safe for him or her or fellow soldiers to be around someone who is consumed with upsetting emotions from the previous night.

It is extremely likely that a soldier will develop a negative belief system. When a Soldier goes into combat, he cannot think to himself, "I probably have nothing to worry about walking down this street, I feel safe." He has to adopt a negative belief set to remain safe. He must think to himself that anyone around him could be a potential threat. He cannot become complacent in his thoughts or behaviors based on the same beliefs the average citizen can have in the United States. The same soldier may resort to anger as his first emotion; it is difficult to fathom, but if a soldier's first reaction to an event is one of understanding or a different positive response, it will be extremely challenging to pull a trigger on a weapon. A negative state of beliefs and mood is his or her best chance for survival in traumatic events. Soldiers have to fight; if they freeze lives are lost.

The quick arousing reaction to events in the environment is the hypervigilant tendencies described in the fourth category of symptoms. A soldier's brain needs to react to every stimulus. Private Jones cannot afford to hear a loud noise and not have an adrenaline rush course through his brain and body. Private Jones does not know if the sound is a gunshot or a car backfiring. He must be prepared for the worst care scenario and respond ready to protect himself and his fellow soldiers.

Symptoms such as hypervigilance, anger, and avoidance of certain places are all vital behaviors repeatedly trained on in the day-to-day life of a military member. The behaviors are viewed as normal and necessary from a society that dictates them to retain a meaning of protection and necessity for survival. The military's way of life creates symbolic interactions of the people, places, objects and events. When the brain and body become fully trained, they will no longer know the difference between combat and training (Skinner, 1953). Their responses to the world as they know it begin to mirror the behaviors described in the DSM-5 category of PTSD, and the trauma they soon undergo in combat only reinforces and ingrains everything they have been taught (Skinner, 1953).

Implications of a Diagnosis

Through his struggle to remain hopeful and resilient despite the trauma he experienced in a concentration camp during World War II, psychiatrist Viktor Frankl wrote, "Suffering ceases to be suffering at the moment it finds a meaning" (Frankl, p. 23). One of the basic pillars of social work is to understand the person by looking at him or her in the context of his or her environment. As we continue to diagnose PTSD in veterans, what more is our country doing than first training and encouraging our soldiers to fight to protect our freedoms and lifestyles and then pathologizing the behaviors that they master to do so. We take away the meaning of the suffering soldiers endure when clinicians quickly categorize their emotional reactions and behaviors. Instead of classifying them into a diagnostic category to capitalize on an insurance payment, we need to change our vocabulary and mindset to a strengths-based perspective of teaching trauma transformation.

"Trauma changes the course of a person's life, following traumatic experiences, each survivor faces the question of how to fit those events, whether a one-time occurrence or an ongoing situation, into new understanding of life's meaning and purpose" (Bussey and Wise, p.1). A veteran will remain a military member the rest of his or her life. "As long as the veteran is seen as alien other, whose changed view of the world is to be treated not as a normal response to extreme violence, but as a sickness from which recovery is to be achieved; it may be difficult for veterans to heal and equally difficult for society to grow" (Shay 2002 as cited in Bragin 2010). They will always recall their training and traumas. As a profession, let social work lead the way in recognizing the sacrifice they have made and helping them facilitate a way to transform their trauma into a meaning for their life. There is nothing post-traumatic about the combat experience, do not seek to get past the trauma but to grow in the trauma.

References

American Psychiatric Association. (2000). *Diagnostic and statistical manual of mental disorders* (4th ed.). Washington, DC: American Psychiatric Association.

American Psychiatric Association. (2013). *Diagnostic and statistical manual of mental disorders* (5th ed.). Washington, DC: American Psychiatric Association.

Analysis of VA Health Care Utilization among Operation Enduring Freedom, Operation Iraqi Freedom, and Operation New Dawn Veterans, from 1st Qtr FY 2002 through 4th Qtr FY 2014. Washington, DC: Author. Retrieved from (URL) http://www.publichealth.va.gov/epidemiology

Blumer, H. (1986). *Symbolic Interactionism: Perspective and Method* (1st ed). London, England: University of California Press.

Bragin, M. Can Anyone Here Know Who I Am? Co-constructing Meaningful Narratives With Combat Veterans. *Clinical Social Work Journal* 38:316–326

Breslau N: The epidemiology of trauma, PTSD, and other post trauma disorders. *Trauma Violence Abuse* 10:198-210, 2009.

Bussey, M. & Wise, J.B. (2007) *Trauma Transformed: An Empowerment Response.* Chichester, NY: Columbia University Press.

Cockerhan, W.C. (2003) *The Military Institution.* In Reynolds, L.T., Herman-Kinney, N.J. (Eds.), *Handbook of Symbolic Internationalism* (pp.491-510). Walnut Creek, CA: Altamira Press.

Frankl, V.E. (1959) *Man's Search for Meaning.* Boston: Beacon Press.

Friedman, M.J. (2013). Finalizing PTSD in DSM-5: Getting here from there and where to go next. *Journal of Traumatic Stress, 26,* 548-556.

Friedman, M. J., Resick, P. A., Bryant, R. A., & Brewin, C. R. (2011). Considering PTSD for DSM-5. *Depression and Anxiety, 28,* 750-769.

Gabbard, G. (2014). *Psychodynamic psychiatry in clinical practice* (5th ed). Washington, D.C.: American Psychiatric Press.

Shay, J. (2002). *Odysseus in America: Combat trauma and the trials of homecoming.* New York: Scribner.

Skinner, B.F. (1953). *Science and Human Behavior.* New York: Macmillan.

Psychosis and the DSM: A Study in Frustration

Andleeb Jawaid

Working as a clinician, one steadily becomes aware of all different types of diagnoses and issues that clients will face—and sometimes it feels as if the clinician will be struggling to find a diagnosis, not for the betterment of the client, but only because the administration requires it. Once one actually starts working in the "field" there is a realization that there are a lot of rules and regulations, and following agency protocol is one of them, and very imperative. In terms of company policy, especially if it is a non-profit, factors such as productivity, budget, funding, and various other things that may not be at the forefront of a clinician's mind begin to take precedence. Unfortunately, this can become quite a conundrum when it comes to diagnosing, as sometimes it is not so much a necessity to understand the client's predicament, but rather a requirement to fulfill billing targets. This issue with billing can be somewhat manageable if the diagnosis at hand is of the more common variety, such as Major Depressive Disorder, etc, as there are more clear cut definitions and characteristics in the DSM -5 itself that can be utilized—however, one diagnosis that can be quite problematic is that of Psychosis NOS—not otherwise specified.

Psychosis itself is not usually an isolated diagnosis, as according to the DSM comes with either Schizophrenia, or will be presented with Depression (APA, 2013). At times, Psychosis can be caused due to substance abuse, known as Substance/Medication-Induced Psychotic Disorder, (APA, 2013), or Psychotic Disorder Due to Another Medical Condition—and all these reasons are justifiable, but the DSM does not take into account *pure* Psychosis, for lack of a better explanation, that may have been exacerbated by traumatic life events. Pure Psychosis is not actually a disorder, but I feel there is a need for there to be a clearer explanation for it, as it can be frustrating not only providing treatment, but also working within a specific framework to meet all requirements.

There is always an inclination to ensure that Psychosis will always be accompanied by another disorder, never on its own, which makes it even more trying, because not only is Psychosis itself difficult for the client to understand, but it can also be even more difficult to provide therapy or arrange treatment plans around such a disorder. Working with an individual who has psychosis feels a bit like falling down the rabbit hole: at times there are instructions, and other times, both the clinician and the client are relying on each other to understand.

Psychosis NOS, which had the diagnosis of 298.90, has now been changed to Unspecified schizophrenia spectrum and other psychotic disorder, which has the same diagnostic code of 298.90, but the definition and description of the disorder is just as vague. It could be argued that even with the DSM 4, the criteria for Psychosis NOS was not very descriptive, as it states "there is inadequate information to make a specific diagnosis or about which there is contradictory information, or disorders with psychotic symptoms that do not meet the criteria for any specific Psychotic Disorder" (APA 2000).

Herein lies the problem—the inadequate information itself lends to an inadequate definition and does not provide enough details when it comes to properly diagnosing a client. The larger issue at hand comes in when there is a quota to meet for adequate billing, for when there are audits and funding is dependent upon matching the numbers—which might be the case in a community counseling setting, with state funding and grants. However, even if one were to forego all these problems, there is also the issue of diagnosing someone with psychosis itself—the criteria and its similarity to some disorders is too closely tied, yet the criteria for other similar disorders, such as Schizophrenia may be too different.

> Unspecified schizophrenia spectrum and other psychotic disorder, previously known as Psychosis NOS, is now defined in the DSM 5 as:
>
> Presentations in which symptoms characteristic of a schizophrenia spectrum and other psychotic disorder that cause clinically significant distress or impairment in social, occupational, or other important areas of functioning predominate but do not meet the full criteria for any of the disorders in the schizophrenia spectrum and other psychotic disorders diagnostic class. The unspecified schizophrenia spectrum and other psychotic disorder category is used in

situations in which the clinician chooses not to specify the reason that the criteria are not met for a specific schizophrenia spectrum and other psychotic disorder, and includes presentations in which there is insufficient information to make a more specific diagnosis (e.g. in emergency room settings) (p. 122).

While this definition may sound concise—it is, but only for a temporary case. If one were to consistently work with a client who has this diagnosis but does not present with any of the other symptoms associated with schizophrenia, is it a sufficient diagnosis? Personally, coming from a clinical standpoint, through the experience of working with such a client, I have to disagree—I do not think that the abovementioned description for Psychosis suffices.

I have worked in a mental health agency for more than two years, and one of my first clients, with whom I am still working currently, was referred from a mental health hospital after being brought into the emergency room by the police. Her discharge report stated the behavior which had led to the hospitalization, and it included paranoid thoughts claiming that her neighbors were plotting against her, that her roommate was trying to steal her identity by stealing her passport and Social Security card, as well as the delusional belief that people (whether her neighbors, her roommate, or anyone she would encounter, such as on the train) could read her thoughts, and in turn would answer back according to whatever was on her mind. My client, who for confidentiality purposes, shall be referred to as Q, initially came with a dual diagnosis of Schizophrenia *and* Psychosis NOS, as noted by the doctors in the hospital.

After undergoing a psychiatric evaluation, Q's diagnosis of Psychosis NOS was maintained, but taking her mental health history and social history into account, the diagnosis of Schizophrenia was removed. As stated in the criteria for Psychosis NOS, which correlated with the diagnosis received at the hospital. As stated in the DSM 5, Unspecified schizophrenia spectrum and other psychotic disorder "includes presentations in which there is insufficient information to make a more specific diagnosis" (APA, 2013), which was applicable when Q first started receiving services.

However, as it has been more than 2 years since Q has been coming in as a client, the diagnosis has not changed—and through consultation with the psychiatrist, and after research and evaluation on my part, it appears that with

13

Q's symptoms, no other diagnosis fits. It can be somewhat confusing when the medication Q is prescribed is used to treat Schizophrenia, yet she is not Schizophrenic—which sounds like a very simple statement, but unless one studies the differences between the diagnoses very diligently, it can be difficult to deduce why one diagnosis is utilized over the other. Q was initially prescribed Aripiprazole, commonly known as Abilify, usually shown in commercials to treat Bipolar disorder or Depression, but after consistently complaining about stomach pain—she was later prescribed Lurasidone, also known as Latuda, another medication used to treat Schizophrenia.

This blurriness between the symptoms of these differing diagnoses can be difficult, especially as one begins treatment, although the most important thing is meet your client where they are at, one of the most upheld tenets of Social Work. However, when your client is out of touch with reality, it can be frustrating, because while you would like to form an empathetic and trusting relationship, you also cannot indulge your client too much, because it is akin to falling down the rabbit hole. This can be frustrating because although Psychotic symptoms such as delusions or hallucinations can occur with Depression or Bipolar Disorder, or more obviously Schizophrenia—there is not enough information for Psychosis NOS. It almost seems like a diagnosis to fall back on, if nothing else fits.

For a beginner-level therapist, having one of your very first clients be so highly symptomatic can be quite an enlightening, educational experience. There is the excitement of finally seeing how theoretical symptoms that have simply been listed in a book actually manifest in a person. Some are so grossly obvious, it is almost as if one is checking off boxes of symptoms observed…whereas with others, there is the humbling experience of realizing that along with Psychosis, there are a handful of other contributing factors which may lead to the issues that the client is facing—and that there is more to the client than just symptoms.

Using Q as an example, her traumatic experiences most likely triggered her psychotic episodes, which have lasted longer than the criteria for Brief Psychotic Disorder, 298.80, which must have at least a presence of one of the following in order to be considered: delusions, hallucinations, disorganized speech, grossly disorganized or catatonic behavior, and the time must be "duration of an episode of the disturbance is at least 1 day, but less than 1 month, with eventual full return to premorbid level of functioning" (APA 2013). My efforts to fully understand Q's diagnosis and whether it is suitable might sound fruitless as the diagnosis was made by the psychiatrist; however, the description for Unspecified Schizophrenia spectrum and other disorders is so dissatisfying that even understanding it is confusing.

For the duration of this paper the diagnosis formerly known Psychosis NOS shall continue to be referred to as that, for the sake of convenience, as it is more concise. The name Unspecified Schizophrenia spectrum and other disorders lacks clarity and is so deliberately abstract that it almost matches the symptoms as they are presented in a client. Psychosis NOS is also so closely related to Schizophrenia, that trying to figure out the difference between the two can only be summed up as one having Catatonia as a symptom, while the other does not.

Viewing Psychosis through the person-in-environment lens, one can draw connections as to why an individual might be triggered by certain events, or how events can have a long-term, traumatic effect on the individual's functioning. To further use the example of Q, events in her life including sexual abuse at the hands of a relative, as well as suffering domestic violence at the hands of her now ex-husband most definitely served as catalysts for her impending psychotic episodes, as such harrowing events left her without any sort of support.

What the DSM fails to provide is any other sort of explanation, especially when it comes to Psychosis NOS, yet there is adequate information on Schizophrenia, and the blurred lines resurface, as the following description can apply to Psychosis, yet when it comes to paperwork and diagnosing, it is one or the other: "Some individuals with psychosis may lack insight or awareness of their disorder (i.e. anosognosia). This lack of 'insight' includes awareness of symptoms of schizophrenia and may be present through the entire course of the illness. Unawareness of the illness is typically a symptom of the illness rather than a coping strategy" (APA 2013). This definition, while incredibly insightful, only serves to confuse, if only on logistical terms—first, this is a very vital piece of information, yet it is included under Schizophrenia. If one were completely ignorant on reading and understanding the DSM, it would be natural to skip over to the criteria for Psychosis NOS, rather than taking the time to read through *all* the correlating disorders.

Secondly, to reiterate a frequently made point, it is quite frustrating to realize that the DSM, a *manual*, does not offer enough information on such a hard disorder, as there are so many distinctions and conditions for each disorder, but some are so subtle that it leaves one wondering if it might be completely futile to diagnose, if it wasn't incumbent on agency, or rather national policies. Although it would only add to a clearer understanding of the disorder if the DSM were to provide more details. Unfortunately it does not—requiring the therapist to investigate on their own. Although at times Q lacks awareness of her symptoms, as stated above, as she, at times, firmly holds on to the belief that there are people who can read minds, and will actively verbalize whatever she is thinking, despite evidence being to the contrary for the clinician. Then—just when it seems that Q's symptoms can neatly be wrapped into a little box of Psychosis NOS and its criteria, she will admit that perhaps there is no such thing as mind-reading, and she only thinks so when she is stressed. For the clinician, it is wonderful if the client begins to show insight into her behavior—but then, what diagnosis matches? What intervention matches? Where does the clinician go from there? As the DSM does not provide answers, the clinician must begin the research process to better explain the client's symptoms and the reasoning behind them.

After gathering history from Q, as well as other collateral sources, it appears Q's symptoms began manifesting themselves after her husband began abusing her, physically, mentally, verbally, emotionally, in any capacity possible. According to research about low self-esteem and fear, by Lincoln et al., it appears "paranoia has been shown to be associated with the perception of low social power and negative social comparisons (Gilbert, Boxall, Cheung, & Irons, 2005). Thus, paranoid beliefs might be linked to the interpersonal self-concept of not being respected or accepted by others" (2010). As Q developed a sense of paranoia due to her husband, it only grew exponentially as she encountered further harsh situations, such as residing in a shelter, as well as coping with a divorce, as well as being an immigrant in general. As she has stated herself, and as evidence shows, she has been left shaken up, with low confidence due to all these harrowing events, and thus her paranoia leads to further delusions—which explains the Psychosis: Q has experienced a deep sense of powerlessness due to the domestic violence, and thus has low confidence and a decreased sense of self-worth as evidenced by her statements and actions.

Working as Q's therapist within an agency setting has been challenging for more than one reason. Some can be cited as billing issues, for example when Q is in a more symptomatic state and is dealt as a Crisis event, and the billing follows accordingly—while times when she seems to be managing her symptoms, have to be adjusted. The other challenge is the overlapping of all the diagnoses. "There are five domains of psychopathology that define psychotic disorders. The level of psychosis, the number of symptoms, and the duration of psychosis are the gradients that have been used to demarcate psychotic disorders from each other and continue to be used for the same purpose in DSM-5" (Heckers 2011), and it is difficult to ascertain the levels at times, especially when the client cannot provide an accurate timeline.

It would be naïve to assume that there would be changes made by the time the next DSM is published, if it is still in use at that time—however, there should be more research into the psychotic disorders, as Schizophrenia seems to be at the forefront, yet there are other diagnoses, but the criteria for Schizophrenia also falls short, because to summarize from the DSM, only two or more of the symptoms listed need to be present, which is so vague and loosely-structured that individuals may be toeing the line between Psychosis and Schizophrenia, which are two very differently perceived disorders, and the stigma attached holds strong reactions and repercussions for the individuals diagnosed, therefore there needs to be more concrete information and research available.

References

American Psychiatric Association. (2000). *Diagnostic and statistical manual of mental disorders: DSM-IV-TR*. Washington, DC: American Psychiatric Association.

American Psychiatric Association., & American Psychiatric Association. (2013).*Diagnostic and statistical manual of mental disorders: DSM-5*. Washington, D.C: American Psychiatric Association

Heckers, S., Barch, D., Bustillo, J., Gaebel, W., Gur, R., Malaspina, D., Carpenter, W. (2013). Structure of the psychotic dis orders classification in DSM-5. *Schizophrenia Research*, 11-14.

Lincoln, T. M., Mehl, S., Ziegler, M., Kesting, M. L., Exner, C., & Rief, W. (January 01, 2010). Is fear of others linked to an uncertain sense of self? The relevance of self-worth, interpersonal self-concepts, and dysfunctional beliefs to paranoia. *Behavior Therapy, 41,*2, 187-97.

Anorexia Nervosa: A Feminist Perspective

Amy Patterson

"Nothing tastes as good as skinny feels." This is the message that super-model Kate Moss promoted after celebrating a milestone in her modeling career. This quote can be found all over the Internet, and women can purchase apparel with the quote on it. With our society idolizing models with skeletal figures, such as Kate Moss, it is not surprising that eating disorders are prevalent in our world today. The media is constantly highlighting the weight loss of celebrities and is quick to criticize anyone who appears to be gaining weight. Body image has become a central focus in our society, particularly for women. The media is constantly conveying a message that you must have a perfect body in order to be accepted and an obsession, with different weight loss trends and diet fads poses a great risk for the development of an eating disorder.

According to the DSM-5 (APA, 2013, p. 338), a person must meet three different criteria in order to be diagnosed with anorexia nervosa. The first criterion is restriction of energy intake relative to requirements, leading to a significantly low body weight in the context of age, sex, developmental trajectory, and physical health (criterion A). The second criterion is intense fear of gaining weight or of becoming fat or persistent behavior that interferes with weight gain, even though at a significantly low weight (criterion B). The final criterion is disturbance in the way in which one's body weight or shape is experienced, undue influence of body weight or shape on self-evaluation, or persistent lack of recognition of the seriousness of the current low body weight (criterion C). The DSM-5 lists several risk factors, including temperamental factors (anxiety disorder, obsessional traits in childhood), environment (culture), and genetic and physiological factors (biological relatives with anorexia, bipolar disorder, depression, or brain abnormalities). By analyzing research on anorexia nervosa as well as current prevention programs in place, it can be argued that adding gender as a potential risk factor could decrease the prevalence of the disorder.

Anorexia is characterized by an obsession with one's body image and a strong desire to be thin. Anorexia is frequently described as a mostly female disorder, and this concept dates back to 1689. Morton described it as a "form of hysteria and a hereditary abnormality of the central neurological system that appears only to young females" (Wozniak, Rekleiti, & Roupa, 2012, p.257). In 1895, Freud associated anorexia with melancholy and sexually immature females. Anorexia is more prevalent today than in past decades, and about 1% of adolescent females in the United States are anorexic. The disorder occurs roughly 10-20 times more frequently in females than in males in developed countries and most anorexic patients have experienced some type of trauma in their lives, including abuse or neglect. Anorexic patients frequently relate these negative experiences to their body image and become ashamed and develop low self-esteem (Wozniak, et al., 2012, p. 258).

McKinlay and Marceau note that "Risk factors and risky behaviors are obviously manifested in individuals, but they are *generated and reinforced* within an ecosocial context and they are strongly related to social position" (2000, p. 297). This means that while anorexia typically develops from an individual's deeply rooted trauma, our society and the media are so obsessed with body image that it worsens these trauma victims' chances of developing anorexia. In regards to prevention, factors such as dieting, negative body image, internalization of the pressure to be thin, and negative affectivity can be seen as "epiphenomena of gender inequity." Yet, if prevention programs recognize gender as a risk factor, they could educate and deter vulnerable females from developing anorexia. These programs could also address substance abuse in regards to vulnerable women with a negative self-image, obsession with plastic surgery, low participation in sports when compared to males, and sexual activity (Piran, 2010, p. 186-187). Teachers of these programs could educate both males and females on risk factors, but differentiate these factors on an individual basis to emphasize their impact. Feminists have always paid close attention to how social systems shape girls' lives, but anorexia prevention has rarely been geared towards how women are impacted by society. Typically, these prevention programs focus on all individuals at the micro level and do not take social environment into account. "Expecting individuals, especially children, to change without a corresponding change in their social environment has raised ethical concerns" (Piran, 2010, p. 186).

As discussed previously, women who are striving for self-control and perfectionism or who have anxiety or depression are at greater risk for anorexia. While the disorder is not fully understood or explainable, research suggests there may be a link between eating disorders and negative images in the media in regards to weight loss and body image (Yom-Tov & Boyd, 2014, p. 196). It has been discovered that models across the world have dangerously low body mass indexes (BMI) of under 18.5 kg/m2. As found in the DSM-5, a BMI of 17 or lower is considered mild anorexia, a BMI of 16-16.99 is moderate, a BMI of 15-15.99 is severe, and a BMI less than 15 is extreme. By encouraging these models to maintain such an unhealthy weight, our society is creating a distorted and unhealthy view of what beauty is. Unsurprisingly, anorexia is more prevalent in Western cultures because the Internet is so popular. As Internet access increases across the world, mental health professionals expect to see a rise in anorexia in other countries. Women, particularly adolescent girls, are especially influenced by these images in the media, which raises their risk for disordered eating. However, even though research has found a correlation between disordered eating and exposure to underweight models and celebrities in the media, it certainly does not guarantee disordered eating.

Current journalist and former mental health nurse Roger Cowell discusses why women choose to idolize these negative role models when being healthy is more important than "looking good." Most models are a size zero, which can trigger many serious health issues, including infertility, osteoporosis, vitamin b12 deficiency, anemia, as well as other mental health issues. Despite these serious consequences, the media portrays a message that if you are skinny you will be happy. When asked to ban models at an unhealthy weight (like size zero), London Fashion Week refused and said "fashion looks good on thin girls" (Allen, 2007, p. 24). Modeling agencies, the media, and other sources of a negative influence on body image need to be held accountable for the promotion of this unhealthy lifestyle that has a variety of very serious, sometimes life threatening risks.

There have been countless stories in the media of celebrities who were idolized by our culture for their body and either died from an eating disorder or were able to recover after nearly dying. Examples of celebrities who are idolized for their skeletal frames today include Kate Moss, Nicole Richie, Mary Kate Olson, and other models. These women are very popular in western media, and their thinness has been frequently associated with sexiness. Another example of a celebrity in the media is Cheryl Burke. She was a former star on a reality television show who was criticized for gaining five pounds by the talk show's hostess, Tyra Banks. This is another reason that the public needs to be educated on the dangers of eating disorders, in particular anorexia nervosa. The average person can see up to 600 advertisements per day, and these are typically filled with our society's misconstrued idea of beauty. The United States advertising industry spends roughly $100 billion per year promoting diet fads and weight loss trends and frequently associates their message with happiness (Curry & Ray, 2010, p. 360-361).

A study conducted in 2012 by Elad Yom-Tov and Danah M. Boyd analyzed Internet searches of 9.2 million users over a 5-month time period on major search engines such as Google and Bing, as well as anorexic activity searches. Anorexic activity searches include searches for pro-ana (pro-anorexia) websites, such as "thinspiration." These sites have blogs for the public to view and post on, tips and tricks on how to become anorexic, as well as pictures of anorexic women that can be used as inspiration to eat less and keep getting skinnier. Obviously there is a lot of controversy behind these websites, and that will be discussed later. This study found that users who searched for anorexic activity were more likely to also search for celebrities who are anorexic or are rumored to have anorexic tendencies, implying that there is a strong association between the two. The study also found that users who searched for anorexic celebrities in the media were roughly two and a half times as likely to search for pro-ana websites compared to other users (Yom-Tov & Boyd, 2014, p. 200-202). Upon review of this study, it may be beneficial for society if the media associated anorexia with celebrities who are at an unhealthy, low body weight and discussed the dangers and risk factors associated with their health. Skinny celebrities frequently get praised and idolized for their skeletal frames, and that is what many adolescent teenage girls are obsessing over.

As earlier discussed, there is a lot of controversy regarding pro-anorexia websites that can be accessed with any computer using search engines. Unless they are blocked by web administrators, these websites can be accessed by any person of any age. While they were initially designed as a safe place for people struggling with anorexia to find support, many of them have a much greater negative influence and message than a positive one, and Karen Dias's article in

2003 discusses these influences. There are anorexic blogs that the public can post to and read, tips and tricks on how to stay skinny or become anorexic, pictures of anorexic women to help motivate users to stick to their anorexic ways, etc. These websites receive a lot of backlash from professionals within the field, even though some suggest they may be necessary for someone struggling with an eating disorder. Dias's article discusses third-wave feminism, which is a movement that is encouraging and creating solidarity among women on the basis of challenging dominant power relations (Dias, 2003, p. 32). Dias highlights the connection and solidarity that pro-ana websites create, which challenges power relations. Her research analyzes a number of different pro-ana websites and discusses the pros and cons of these sites and how they may be helping people who struggle with anorexia. While some of the negative aspects of these websites are listed above, the positive aspects will be discussed next.

One important aspect of these websites is that even though they can be accessed by the general public, they typically have warning messages on their homepage that state that they are specifically designed for women who already have anorexia nervosa, not women who are trying to become anorexic. However, many sites also state that they are not intended for people who wish to go into recovery, but simply those who are content with their anorexia. Users of these websites frequently post pictures of themselves in revealing clothes to get support from other anorexics, but they usually are insulted for their weight and told they need to continue getting thinner. In 2001, the Anorexia Nervosa and Associated Disorders advocacy group reached out to major search engines, such as Yahoo and Google, to take down these websites. Within just a few days, over 100 sites were shut down. However, these websites are nearly impossible to follow consistently, as they are frequently shut down and are created under a different URL within a matter of days.

There are many stories about women who have nearly died after becoming obsessed with pro-ana websites, and an extreme case is that of Grainne Binns. Grainne is a teenage girl who almost died of anorexia nervosa in part by an obsession with anorexia blogs. Grainne Binns became addicted to posting pictures of her skeletal frame on pro-ana sites and constantly checked the comments that other users left. Despite her extremely low weight, people on the site would continually tell her that she was fat and needed to lose more weight. After about a year, she had lost almost a third of her bodyweight with a dangerously low BMI of 14, which according to the DSM-5 is a case of extreme anorexia, the most dangerous type. After 10 weeks of intensive inpatient hospital treatment, she still kept with her unhealthy diet and continued to starve herself. It was not until later when her family sat her down and explained their worry and concern with her body weight that she finally turned things around. While this is an extreme example of the dangerous affects a pro-ana website may have, it shows just how serious they can be and how they can impact someone to nearly dying.

Despite all of the negative aspects of these websites, Dias's research on anorexia survivors suggests that there may be some positive outcomes associated with pro-ana websites. Studies of recovery from anorexia frequently focus on a medical model that assumes professional intervention is necessary; however, many recovered anorexics claim that people and events outside of therapy were what ultimately helped them recover.

Narrative therapy, specifically the "externalization of the 'eating disorder voice'" (Dias, 2003, p. 39), is a practice that clients are sometimes encouraged to practice. This allows the client to isolate the voice that encourages this unhealthy behavior and distance themselves from it. It may help the client to realize that the eating disorder does not define them and may be done in the form of letter writing. Pro-ana sites sometimes provide a safe space for anorexic people to voice their hatred of this disorder and can help them try to gain control over it. Examples of positive aspects of pro-ana websites show that some users are using these websites as they work towards recovery and help support others in similar situations. Dias concludes by stating that these websites need to be looked at in a different light in order to maximize their benefits for those who are on the road to recovery from anorexia and to be understood in the context of the new, third-wave feminism.

Taking into consideration the pros and cons of pro-ana websites and other messages and images we are exposed to by the media, education and prevention of anorexia nervosa play a key role in the disorder. Looking at the disorder from a feminist standpoint, it would be most beneficial to our society to add gender as a risk factor for developing anorexia nervosa along with other risk factors we are exposed to through cultural ideations in our social environment. Research shows that anorexia nervosa is more prevalent in women than men. The constant pressures from the media force women to feel the need to be extremely thin in order to be happy and this is very detrimental, especially in regards to adolescent females who are at higher risk for becoming anorexic. Educating women on the dangers of the low body weights of celebrities and models can help prevent rates of anorexia from rising in the decades to come.

References

Curry, J., & Ray, S. (2010). Starving for Support: How Women With Anorexia Receive 'Thinspiration' on the Internet. *Journal of Creativity in Mental Health, 5*(4), 358-373. Retrieved from http://web.ebscohost.com/ehost/detail/detail?sid=330f6d68-0180-4bbb-9a63-ea7e43
9d69f9%40sessionmgr4004&vid=22&hid=4101&bdata=JnNpdGU9ZWhvc3Qtb
Gl2ZQ%3d%3d#db=rzh&AN=2010884719

Dias, Karen (2003). The Ana Sanctuary: Women's Pro-Anorexia Narratives in Cyberspace. *Journal of International Women's Studies*, 4(2), 31-45.

Macsween, M. (1995). *Anorexic Bodies: A Feminist and Sociological Perspective on Anorexia Nervosa*. New York, New York: Routledge.

Norris, M. L., Boydell, K. M., Pinhas, L. and Katzman, D. K. (2006), Ana and the Internet: A review of pro-anorexia websites. Int. J. Eat. Disord., 39: 443–447. doi: 10.1002/eat.20305

Piran, N. (2010). A Feminist Perspective on Risk Factor Research and on the Prevention of Eating Disorders. *Eating Disorders, 18*(3), 183-198. Retrieved from http://web.ebscohost.com/ehost/pdfviewer/pdfviewer?sid=330f6d68-0180-4bbb-9a63-ea7e439d69f9@sessionmgr4004&vid=3&hid=4101

Tanton, E., Cowell, R., Houghton, G., & Scullion, J. (2007). Readers panel. A recipe for eating disorders. *Nursing Standard, 21*(28), 24-25.

Yom-Tov, E., & Boyd, D. (2014). On the link between media coverage of anorexia and pro-anorexic practices on the web. *International Journal of Eating Disorders, 47*(2), 196-202. Retrieved from http://web.ebscohost.com/ehost/pdfviewer/pdfviewer?sid=330f6d68-0180-4bbb-9a63-ea7e439d69f9@sessionmgr4004&vid=20&hid=4101

Zucker, N., Wagner, R., Merwin, R., Bulik, C., Moskovich, A., Keeling, L., & Hoyle, R. (2015). Self-Focused Attention in Anorexia Nervosa. *International Journal of Eating Disorders, 48*(1), 9-14. Retrieved from http://ejournals.ebsco.com/Direct.asp?AccessToken=3P1X1XD81QS0LL2NZ-SXLX-P2ET-8SXLL&Show=Object

Kleptomania: The Stereotypical Psychiatric Ailment of the Female Population

Courtney Chambers

Introduction

The term kleptomania, derived from the Greek meaning "stealing madness," appeared first in the early 19th century (Murray, 1991). Thefts by a Kleptomaniac are usually unpremeditated and resemble classic compulsion. According to the *Diagnostic and Statistical Manual of Mental Disorders (DSM-5)*, the diagnostic criterion for kleptomania 312.32 is a "recurrent failure to resist impulses to steal object that are not needed for personal use or for their monetary value." There is an "increase in the sense of tension immediately before committing the theft and pleasure, gratification or relief at the time of committing the theft." It is made clear within the criteria that the "stealing is not committed to express anger or vengeance and is not in response to a delusion or hallucination," nor is the disorder "better explained by conduct disorder, a manic episode, or antisocial personality disorder" (American Psychiatric Association, 2013).

According to the DSM-5, "Kleptomania occurs in about 4-24% of individuals arrested for shoplifting" (American Psychiatric Association, 2013). Kleptomania is a highly rare disorder occurring in only 0.3%-0.6% of the population. It is one of psychiatry's most poorly understood diagnoses and typically goes undiagnosed in clinical settings. The disorder begins in adolescence and if left untreated, can become a chronic illness (American Psychiatric Association, 2013).

Intellectual History

Prior research on kleptomania, beginning in the early 1800s, implemented the definition as "uncontrollable and irrational stealing" (Fullerton, 2003). Other researchers have described Kleptomania as, "a unique madness characterized by the tendency to steal without motives and without necessity" (Fullerton, 2003). Beginning in the late 1800s and early 1900s, Kleptomania was named "a

---

conscious urge to steal occurring in an individual in whom there is no ordinary disturbance of consciousness. The individual frequently strives against this urge, but by its nature it is irresistible" (Fullerton, 2003). As the decades progressed, the causes of Kleptomania have remained unknown. However, one aspect of kleptomania psychiatrists never disagreed on is the definition. On the contrary, psychiatrists fail to agree on any explanation of what causes kleptomaniac behavior.

One of the earliest essential characteristics of Kleptomania was that the choice of the item made no rational sense. The item was useless to the thief, and it was usually cheap or used (Fullerton, 2003). However, first generation psychoanalyst Sigmund Freud "aspired to explain the underlying dynamics of human behavior" and persisted to develop an in depth psychology of Kleptomania (Fullerton, 2003, pg. 205). Freud gained an understanding of kleptomania and proposed the impulses were associated with sexuality. After Freud's death, psychoanalyst Wilhelm Stekel followed his suggestion to implement a theory based on Kleptomania being "driven by suppressed sexual urges to take hold of something forbidden, secretly" (Fullerton, 2003, pg. 205). Stekel concluded that Kleptomania was a sexual desire suppressed by taking hold of a symbolic object, which was an expression of a mental state.

Stekel's research offered an explanation of why Kleptomaniacs stole ridiculous objects and how they fantasized the object as a sexual urge. However, psychoanalyst Alfred Alder opposed Stekel's idea and offered a different perspective. "To Alder, the strong unconscious urges that drove neurotic behavior such as Kleptomania arose from the neurotic's deep feelings of physical and social inferiority rather than from sexual urges and frustrations" (Fullerton, 2003, pg. 206). Alder arranged Kleptomania as a way to combat one's feelings of inferiority through the acts of shoplifting rather than to gain sexual pleasure through stealing. Adler's concept of the inferiority complex gave an interesting alternative view as a proposed theoretical basis of kleptomania, but lacked social research in proving his theory.

Psychoanalyst Fritz Wittels argued that, "Kleptomaniacs were sexually underdeveloped people who felt deprived of love and had little experience with human sexual relationships" (Fullerton, 2003, pg. 206). Wittels used his research to prove that stealing was the sex life of a Kleptomaniac. Wittels's study expounded on his proposal that kleptomaniacs were sexually underdeveloped, by stating that the sexual desires in men and women were influenced by stealing. Similar to the research of Stekel, Wittles also developed a theory, using kleptomania as a sexual influence on a person. Wittles argued that the sexual inhibitions of individuals lay in the powerful thrill of shoplifting. Contrary to Wittles, Franz Alexander, Otto Fenichel and Sandor Rado examined Kleptomania through a representation of the losses in childhood rather than through an explicit sexual connotation.

Alexander, Fenichel and Rado developed what are known as, the castration complex, the Oedipus complex and the penis envy. Through the complexes, the definitive basis of Kleptomania lies within early childhood development. The Kleptomaniac uses the items stolen, to symbolically represent the losses of childhood. For example, leaving the mother's womb, being weaned from breast feeding and giving up feces in toilet training can each configure a loss for a Kleptomaniac, and in their minds, they have a right to get them back (Fullerton, 2003).

Some of the stolen items that psychoanalysts found symbolically meaningful to kleptomaniacs included anything that reminded them of feces, a mother's milk, a penis and sometimes even the mother's body. According to the psychoanalysts, kleptomaniacs would steal objects that represented these losses. Alexander suggested that the "unconscious memories of the shocks of birth, weaning, and toilet training can remain as a generalized anxiety about future shocks. Compulsive shoplifting was a way of combating weakness and vulnerability" (Fullerton, 2003, pg. 206).

Although past psychoanalysts depicted the cause of Kleptomania and sought to pinpoint the drive behind it, the cause has yet to be defined. According to a review of Kleptomania, current psychoanalytic theories link "compulsive stealing to childhood trauma and neglectful or abusive parents, and stealing may symbolize repossessing the losses of childhood" (Murray, 1991, pg. 133). The disorder is highly associated with mood disorders, depression and anxiety spectrum disorders. Kleptomania is also regarded as, "a form of addictive behavior and has been shown to be associated with other substance use disorders" (Grant, 2010, pg. 292). Newer studies of Kleptomania have now been reported to be caused by head trauma, and that the traumatic brain injury initiated the Kleptomania disorder. Although many researchers have made attempts to theoretically grasp the underlying concept of Kleptomania, much progress has yet to be made. "Over time, there have been serious disagreements on how many true Kleptomaniacs actually exist and on what contributes to Kleptomania" (Fullerton, 2003, pg. 207).

Although a diagnosis of Kleptomania is a rare and highly misunderstood disorder, its commonalities with substance use and addictive disorders remain supported through research. As the DSM points out, psychiatric comorbidity of other disorders being associated with Kleptomania is quite common. In specific, substance use disorders, eating disorders, depressive disorders, anxiety disorders and personality disorders are most commonly associated with kleptomania (American Psychiatric Association, 2013). Like individuals with substance use disorders, a kleptomaniac "has tension prior to stealing and experiences relief or gratification after stealing" (Grant, 2010, pg. 291). The tension Kleptomaniacs experience right before the theft is similar to that of a substance use disorder, where a craving is emphasized at its ultimate high.

The compulsion Kleptomaniacs experience right before committing a theft is an irresistible state, and when attempting to stop the behavior, Kleptomaniacs may exhibit withdrawal symptoms. Similar to substance use disorder, Kleptomaniacs may experience "insomnia, agitation, and irritability" (Grant, 2010, pg. 293) when attempting to stop their behavior. Kleptomaniacs also report the need to steal "more expensive or riskier items during the course of their illness in order to achieve the same "rush" they initially felt" (Grant, 2010, pg. 293), which is a similar trait of those diagnosed with a substance use disorder.

Although the value of an object is of less importance to a Kleptomaniac, the impulse is greatly impacted when the thrill is no longer apparent. Kleptomaniacs then begin to give their thefts a criterion and gain a preference for items that have a larger value.

Kleptomania then begins to exhibit similar traits to those of a compulsive shoplifter, where much value is placed on the object. As a Kleptomaniac develops a criterion for the items that are stolen, how then is being a Kleptomaniac measured? When the motivation behind the theft is deliberated by the usefulness of the object, would that person then be named a compulsive thief versus a Kleptomaniac? Would a Kleptomaniac then exceed the boundaries and criteria for the diagnosis and be labeled as a common thief rather than exhibiting a rare disorder? If the thefts are then deliberated, or thought out before hand, one is no longer experiencing impulsive symptoms, but rather compulsive symptoms.

Feminist Theory

According to the *Theoretical Perspectives for Direct Social Work Practice*, "analyzing the varied experiences of people from political perspective that holds a sex-based analysis as one of the key analytical lenses constitutes a feminist approach" (Coady, 2008, pg. 344). The feminist theory is a gender-based framework that gives women a sex-based perspective in understanding the social limitations women are faced with every day. The feminist theory seeks to empower women by promoting the social, political and economic quality of the sexes. One of the main goals of Feminism is to give women the importance they deserve and reveal that women have historically been subordinate to men (Coady, 2008).

Feminism provides a "critical examination of the individual and collective choices that shape women's lives," (Charter, 2015, pg. 73). Concisely, "feminism is the struggle to end sexist oppression" (Charter, 2015, pg. 74). Ending the sexist oppression would then constitute a society where men and women can be seen as equals and are less likely to be controlled by gender stereotypes. The feminist theory also illustrates the issues that affect women and enlightens society on how the issues have evolved over the years (Charter, 2015).

Diagnostic Critique

Kleptomania is a convenient stereotypical psychiatric ailment that is more widespread than acknowledged and disproportionately afflicts the female population. According to the DSM-5, Kleptomania is more prevalent in females than in males, at a ratio of 3:1 (American Psychiatric Association, 2013). It is proposed that women are more likely to steal because of their shopping habits. It is also suggested that women are more likely to be sent to a psychiatric hospital rather than jail after committing a theft. However, kleptomania is exclusively measured from an individual perspective, rather than with psychiatric evaluation.

After committing a crime, an officer is the first to assess the individual and make suggestions for the appropriate punishment for the actual crime. It is through the onsite officer that a decision is then made for a shoplifter to be sent either to jail or a psychiatric facility. As the officer makes his decision, of the most appropriate outcome for the crime, the fate of someone's life then lies in the hands of an amateur psychologist. Thus, the qualifications for kleptomania can be inconsistent and can vary from each individual's perception.

As Fullerton stated in *The Journal of Psychology*, "French psychiatrists more and more concluded that Kleptomania was more a legal excuse for self-indulgent behavior in ladies, than a valid psychiatric aliment" (2003). The illness of kleptomania should not be considered a psychiatric illness if in fact, it is rationalized through the verbal content of the shoplifter herself. As anyone is capable of demonstrating verbal explanation of a mental illness with proper preparation, kleptomania is then perpetuated as an unreal disorder. Through research, it was declared that "one in five apprehended shoplifters was a psychiatric case, and few exhibited all the symptoms of Kleptomania" (Fullerton, 2003, pg. 207). As a rare disorder, Kleptomania can be questionable as to whether or not it deserves a place in the DSM, and should be magnified on rather or not its use can be adequately measured.

Kleptomania disproportionately afflicts the female population and stereotypes the diagnoses of Kleptomania as a female dominated disorder. As stated before, male shoplifters are extremely rare, and those who are apprehended are usually imprisoned. As female shoplifters are arrested, they automatically qualify as being mentally ill. Thus, the female shoplifters are directed to a psychiatric facility, rather than a penitentiary. It may appear as an easy solution for women to complete sentencing in a facility, as opposed to being held in detention. However, it is a convenient diagnosis that enhances male societal power and that radical feminists argue against.

Radical feminists "argue that individual women's experiences of injustice and the miseries that women think of as personal problems are actually political issues" (Coady, 2008, pg. 349). This radical argument is based primarily on making private issues publicized, in order to minimize the reality of male power systems. Historically, "families have been organized according to male lines of inheritance and dependence" and society has also been "constructed in a way that accrues disproportionate share of power to men" (Coady, 2008, pg. 350). The goal of radical feminists is to alleviate the power of sexism, which gives men power over women, and create an equal balance of oppression in the social system.

The mental diagnosis of Kleptomania in women is a convenient label, as women are already viewed as marginalized and disadvantaged in socio-economic status. Liddell and Martinovic state that, "There is a view among theorists and researchers that women's offending is linked to their economic marginality" (2013, pg. 128). Liddell and Martinovic also "contend that women's crime is deeply affected by women's place in society" and is disregarded as having little opportunity for education and the development of job skills (2013, pg. 132).

Socialist feminism "draws on the theories of Karl Marx to explain how economic or material conditions form the root of culture, social organization, and consciousness itself" (Coady, 2008, pg. 347). Socialist feminism primarily focuses on a feminist-based class and seeks to end women's oppression by eliminating capitalism (Coady, 2008). Some aspects of gender specific oppression that are emphasized in social feminism include problems of sexual abuse, insufficient childcare, and constraints on reproductive rights (Coady, 2008). These aspects each have a greater impact on women than men, and reinforce the idea that gender plays a major role in society, in ways that are harmful to women.

In an unequal society, where women are viewed as a lesser "power" than men, a diagnosis of Kleptomania constrains women from having the capability of abolishing the stereotypical interpretation. As the feminist theory suggests, "the barriers of the full potential of women must be challenged and changed" (Charter, 2015, pg. 73). This challenge can be explained by liberal feminism, which "describes society as being composed of separate individuals, each competing for a fair share of resources" (Coady, 2008, pg. 346). Liberal feminism suggests the idea that "society violates the value of equal rights in its treatment of women" (Coady, 2008, pg. 347), by restricting women as a group rather than treating women as individuals in society. Liberal feminists simply argue that women should have the same rights as men including equal education, equal employment opportunities, and equal pay for work (Coady, 2008). Liberal feminists have a goal of prioritizing women and integrating them into society has a whole.

Liberal, socialist and radical feminism are a few of the perspectives that seek to illuminate the disadvantages of women in society through the feminist theory (Coady, 2008). The projected labeling of women as Kleptomaniacs is a problematic issue for the equality and liberation of women in society. The diagnosis of Kleptomania is not only convenient, but it seeks to decrease the global empowerment of women. If the disorder cannot be measured appropriately, with precise psychiatric assessment, then it in fact should not be a diagnosis at all.

References

American Psychiatric Association. (2013). *Diagnostic and statistical manual of mental disorders* (5th ed.). Washington, DC: American Psychiatric Association.

Baxter, J & Taylor, M. (2014). Measuring the socioeconomic status of women across the life course. *The Institute of Family Studies 95*(1), 62-75.

Charter, L. M. (2015). Feminist self-identification. *Journal of Social Work Education, 51:* 72-89. DOI: 10.1080/10437797.2015. 977162

Coady, N., & Lehmann, P. (Eds). (2008). *Theoretical perspectives for direct social work practice.* 2nd ed. New York, NY:Springer.

Fullerton, A. R. (2003). Kleptomania: A brief intellectual history. *The Journal of Psychology 103*(3), 201-209.

Grant, E. J, Odlaug, L. B & Kim, W. S. (2010). Kleptomania: clinical characteristics and relationship to substance use disorder. *The American Journal of Drug and Alcohol Abuse 36*, 291-295. DOI: 10.3109/00952991003721100

Liddell, M & Martinovic, M. (2013). Women's offending: trends, issues and theoretical explanations. *The Journal of Social Inquiry 6*: 127-142.

Murray, B. J. (1991). Kleptomania: A review of the research. *The Journal of Psychology 126*(2), 131-138.

A Place for Everything – Everything in its Place: A Functionalist Perspective of Borderline Personality Disorder

Candyss N. Newman

Understanding over 40 years of exploring borderline personality disorder (BPD) teases out several interesting diagnostic trends, one being that prevalence of the condition increases as the sample population narrows from general to highly specialized clinical settings. According to current research, BPD exists within an average of 1.6% of the general population, roughly 10% of outpatient mental health clinics, and in as much as 20% of the population at inpatient psychiatric facilities (American Psychological Association [APA], 2013; Gabbard, 2014). Another diagnostic development of interest is that, while more current research points to an equal prevalence of this dis-order between men and women, the Diagnostic and Statistical Manual of Mental Disorders historically reports that females represent a much higher percentage of BPD patients than the opposite sex (Gunderson, 2009; Sansone & Sansone, 2011; APA 2013).

Today, the DSM-V states that the estimated rate of BPD diagnosis in women is 75% of those with the disorder (APA, 2013). Some scholars believe this predominately female trend is a result of sampling bias, where studies purporting a gender difference in diagnosis are more a reflection of female over-representation within the particular setting of the population being examined (Sansone & Sansone, 2011). The same researchers attribute this sample setting discrepancy to evidence found stating that men tend to seek treatment in rehab facilities for substance abuse while women utilize treatment within mental health clinics, which are settings typically examined within these studies of prevalence. Diving deeper into these noteworthy trends also signals cause for alarm as the DSM-V specifies that histrionic, borderline, and dependent personality disorders are diagnosed more so in women than men, along with a warning for clinicians to pay closer attention to such biases (APA, 2013). Even though a dis-proportionate prevalence of this personality disorder may be due to differences in gender behavior, under-, over-, and misdiagnosis can also be the unfortunate result of social stereotypes (APA, 2013). Regardless of the justification, this information highlights the tendency to subjectively label individuals' behavior as disordered based on a very broad set of criteria. This chapter is a call to explore the diagnosis of borderline personality disorder beyond the typical perspectives that focus on a person's emotional dysfunction in relation to others, in order to examine how the diagnosis itself functions within society.

Brief History: 'Borderline' as a Personality Disorder

Psychiatry in the first half of the 20th century is known to have been dominated by the psychoanalytic school of thought and for having a system of disorder classification still very much in its infancy. Psychoanalysts during this time were chiefly concerned with categorizing either treatable neurotic patients or untreatable psychotic ones (Gunderson, 2011; Gabbard, 2014). Gabbard's and Gunderson's work show that the term "borderline" surfaces from psychiatrists who identify schizophrenic patients with the propensity to fall in between the quasi-dichotomous and primitive classification system, exhibiting dysfunctional ego identity, impulse-control, and thought-processes. Colloquially, these patients are called borderline when clinicians observe that they do "not fit well into the existing diagnostic rubrics" (Gabbard, 2014). From that point, the borderline illness gained more recognition in the field as a clinically significant syndrome of instability as psychiatrists worked to carve out its veritable array of behavioral features displayed by patients (Gunderson, 2011; Gabbard, 2014). Gunderson's (2011) account of this disorder also outlines the criticisms of this newfound pseudo-syndrome and its champions. He brings to light, even during those early days, how challenges to the psychoanalytical paradigm for treating mental illness are resolved by simply revising the existing psychiatric terminology.

After officially entering the DSM-III by 1980, borderline person-ality disorder moves further away from the diagnostic category of schizophrenia as more personality disorders continue to be defined and psychoanalysis gives way to biologically based psychiatry. Con-currently, growing discourse concerning BPD patients takes a negative turn as clinicians develop a "highly pejorative meaning for the borderline diagnosis" (Gunderson, 2011). More often, these individuals diagnosed with BPD are seen as "fickle, egocentric, irresponsible, love-intoxicated" and "intractable, unruly" patients (Gunderson, 2011, p. 3). An insurgence of BPD research in the 1980's contributes to its distinction from other mental illnesses, namely depression and PTSD, despite the presence of overlapping diagnostic features. Contrary to these positive gains, there is also rising disdain for the DSM-III classification of BPD in the eyes of feminists who view it as a way to pathologize women and blame them for being victims of a history of abuse (Gunderson, 2011). Exploration in the 1990's supports feminist concerns as Becker & Lamb's (1994) study of sex bias in diagnosing BPD and PTSD reveals a propensity for clinicians to view female cases as "more borderline than male cases" (p. 58). Though this is a notable point of interest, the critique of BPD diagnoses in this chapter is not one of men versus women, as functionalism is devoid of a feminist agenda. Additional epidemiological, etiological, and treatment-related investigation over the last 20 years is responsible for the current shape of borderline personality disorder within the DSM-V, which is detailed as follows:

Current Diagnostic Features and Criteria: Borderline Personality Disorder (BPD) is characterized by a "pervasive pattern of instability of interpersonal relationships, self-image, and affects marked by impulsivity that begins by early adulthood and is present in a variety of contexts" (APA, 2013, p. 663). Five or more of the following indicates a diagnosis of BPD:

1. Frantic efforts to avoid real or imagined abandonment.

2. A pattern of intense and unstable interpersonal relation-ships characterized by alternating between extreme ideation and devaluation.

3. Identity disturbance: markedly and persistently unstable self-image or sense of self.

4. Potentially self-damaging impulsivity in at least two areas (i.e. sex, spending, reckless driving, substance abuse, binge eating).

5. Recurrent suicidal behavior, gestures, threats, or self-mutilation

6. Affective instability due to marked mood reactivity (i.e. intense episodic dysphoria, irritability, or anxiety usually lasting a few hours and rarely more than a few days).

7. Chronic feelings of emptiness.

8. Inappropriate, intense anger, or difficulty controlling anger.

9. Transient, stress-related paranoid ideation or severe dissociative symptoms.

Those clients who seek help are often treated using techniques such as dialectal behavior therapy, transference-focused therapy, and mentalization-based therapy (Gabbard, 2014). Likewise, these pa-tients may benefit from greater social adjustment, decreased suicidality and anxiety, as well as increased mood and self-identity stabilization according to contemporary research (Bateman & Fonagy, 2010; Gabbard, 2014). Still, the development of the term "borderline" into a full fledged personality disorder shows glaring details regarding how its diagnosis functions in society beyond the aforementioned benefits to those individuals who receive it.

Functionalism and Social Institutions

The evolution of functionalism, also known as functionalist theory or functionalist perspective, comes from the explorations of various sociological heavyweights since the time of the Enlightenment of 18th and 19th century Europe. (Anderson & Taylor, 2013). As American culture grows of greater interest to those in the field over time, U.S. thinkers nudge sociology from abstract theory and closer to more practical applications. French-born Emile Durkheim is a founding theorist of the functionalist framework whose work chiefly surrounds the constancy and order within society. Likewise, Anderson & Taylor (2013) explain that a functionalist perspective examines the individual aspects that contribute to the whole of society.

According to Durkheim, the framework of functionalism conceptualizes society as more than just the sum of its parts, where each of those components serves a societal "function" or purpose. It places focus on the organization and stability of society that is promoted by the existence of collective social agreements and values (Anderson & Taylor, 2013). Most importantly, according to functionalist perspective, the dysfunction of one feature of society will produce subsequent social problems. In turn, those problems impact all other parts of society. Regardless of whether they are positive or negative in nature, these disruptions interfere with societal balance such that change is required in order to return to the status quo (Anderson & Taylor, 2013). In a nutshell, functionalists believe that society is an ethereal structure made up of shared expectations – social institutions – and relationships governing how groups are supposed to behave as a means of maintaining order. Hiccups, typically set apart as deviant behavior, occur when institutions fail to continue meeting a societal need and lead to the dysfunction of particular social systems that must be adjusted as a means of maintaining equilibrium. Order is established by social institutions' collective agreement on public ideals. This is a theory maintaining the position that there is are parts of our social systems that are fixed as the norm, where each component must remain in place to avoid compromising the strength of the entire structure. As such, inequality is an inevitable and functional part of society (Anderson & Taylor, 2013).

Social institutions, as defined by sociologists in pre-WWII America, are subsystems that govern the expected behavior within society (Newman, 1977; Anderson, 2013). These institutions were largely understood to include family, education, religion, economy, law, and marriage. Functionalists view these parts of society as "established and organized systems of social behavior with a particular and recognized purpose" (Anderson & Taylor, 2013, p. 18). Social institutions are man-made constructs seeded deeply within the cultural history that serve to diminish uncertainty and the costs of social conduct (Branisa, Klasen, & Zeigler, 2013). Because they guide human inter-actions, social institutions are thought to evolve as a product of adaptive behavior (Branisa et al., 2013; Urpelainen, 2011). As society reaches consensus on these organized behavioral systems, a conceptualization of deviance takes shape. The product of developing an understanding of what good behavior entails is the identification of those behaviors that are, by contrast, unacceptable (Anderson & Taylor, 2013). Essentially, social institutions are the systems that maintain societal order through the collective contract that requires its members do what is expected of them according to cultural norms.

Functionalist Perspective of a BPD Diagnosis

Researchers often approach BPD diagnosis with a micro-level functionalist perspective by exploring how the characteristic emotion-al disturbances disrupt the harmony of person-to-person social interactions. Work by Keltner and Kring (1998) posits that this disruption to social institutions plays a role in the maintenance of borderline personality disorder, in that the symptomatic emotional and conduct disturbances contribute to the systematic breakdown of social interactions that those who are diagnosed with the disorder often experience during their lifetime (Keltner, 1998). Accordingly, researchers believe that the emotional dysregulation shown by patients with BPD provides im-

portant information regarding their own behavior and their interaction partners, as well as the way patients approach future social interactions (Keltner &
Kring, 1998). Undoubtedly, as the shape of BPD as a diagnosis progresses over
time, the mental health field identifies symptoms experienced by a population
of people that can be beneficial in clarifying humans' social interactions by facilitating the understanding how others are affected by their undesirable
behavior. However, it is undeniable that controversy exists in determining exactly what constitutes "undesirable".

This chapter critiques the diagnosis of borderline personality dis-order
from its macroscopic function in our society. Through the lens of functionalism, this diagnosis acts as a crucial tool for restoring the "natural" order of
psychiatry as a social institution along with the mainstream social structure of
our country. According to the American Psychiatric Association (2013), the
DSM-V is "a classification of mental disorders with associated criteria designed
to facilitate more reliable diagnoses of these disorders" and "has become a
standard reference for clinical practice in the mental health field" (p. xli). Furthermore, this text provides criteria that are intended to assist trained clinicians
with making objective assessments of a patient's symptomatology. The DSM-V
is a guide for professionals who are associated with the mental health field to
use as they seek out "a common language to communicate the essential characteristics of mental disorders presented by their patients" (APA, 2013, p. xli).
The validity of this text is directly dependent upon the psychiatric community's shared belief in its utility. Moreover, the original concept of "borderline" evolves from clinicians' need to pathologize behavior as a means of
restoring order to the psychoanalytic framework, as pat-terns of patients'
symptoms failed to fit squarely into either neurosis or psychosis (Gunderson,
2009; Gabbard, 2014). As a result, psycho-analysts actively identify wide ranges
of behaviors and come to a collective agreement on their pathological nature.
The DSM-V (2014) speaks to the difficulty in differentiating a cross-sectional
presentation of BPD from an episode of either depressive or bipolar disorder
when clinicians don't have an accurate documentation of the patient's history
or behavioral pattern. Essentially, clinicians and paraprofessionals now have a
catchall category that reinforces the validity of diagnostics as the central focus
in mental health treatment. The question also remains that, as clients seek
treatment for emotional dysregulation and other symptoms that affect their

interpersonal relationships, is it helpful for them to be categorized using criteria that so closely mirror other conditions that it is difficult for trained professionals to make an accurate diagnosis? Additionally, existing research extends the period of human development concerning regulation of impulsivity and decision-making past the age of 18 and into early adulthood, though it was previously thought to conclude in the late teens (MacArthur Foundation Research Network on Adolescent Development and Juvenile Justice, 2006). In other words, the progression of brain and psychosocial maturity for early adults coincides with the peak years of impairment for BPD patients. Knowing that the DSM-V warns against attributing behavior that is related to a developmental phase to a mental disorder, how is the text's classification of BPD to serve as the reliable reference for professionals that it claims to be?

In a culture so focused on the identification of acceptable behavior by way of labeling other actions as deviant, the existence of this diagnosis marginalizes individuals under the guise of medical necessity. As the flaws in using this diagnosis are tallied, more light is shed on the self-serving function of BPD as a justification for behavior that is otherwise unable to associate directly with any other mental disorder. In one respect, identity and emotional instability characterized by BPD helps the whole of society further vilify those who disrupt the social interactions of its members. In another, this diagnosis functions in justifying the legitimacy of mental health clinicians' aptitude for classifying disordered behavior, regardless of whether or not it is expressly warranted. For either instance, it is evident that there is a place for these extreme and unstable individuals who plague mainstream society with their irritable mood swings and inappropriate outbursts of anger. They pose too great a challenge to the status quo of our modern society. Thus, necessary adjustments are made to direct these individuals to their rightful place as psychologically abnormal in order to justify any assertion of dominance over them, thereby bringing society back to equilibrium.

References

American Psychiatric Association. (2013). Diagnostic and statistical manual of mental disorders (5th ed.). Washington, DC: American Psychiatric Association.

Anderson, M.A., & Taylor, H.F. (2013). Sociology: The essentials (7th ed.). Bel-mont, CA: Wadsworth Cengage Learning.

Bateman, A, & Fonagy, P. (2010). Mentalization based treatment for borderline personality disorder. World Psychiatry, 9(1), 11-15.

Becker, D., & Lamb, S. (1994). Sex bias in the diagnosis of borderline personality disorder and posttraumatic stress disorder. Professional Psychology: Research and Practice, 25(1). doi: 0735-7028/94/S3.00.

Branisa, B., Klansen, S., & Zeigler, M. (2013). Gender inequality in social institu-tions and gendered development outcomes. World Development, 45, 252-268.

Gabbard, G. (2014). Psychodynamic psychiatry in clinical practice (5th ed). Washington, D.C.: American Psychiatric Press.

Gunderson, J.G. (2009). Borderline personality disorder: Ontogeny of a diagnosis. The American Journal of Psychiatry, 166(5), 530–539. doi:10.1176/appi.ajp.2009.08121825.

Keltner, D., & Kring, A.M. (1998). Emotion, social function, and psychopathology. Review of General Psychology, 2(3), 320-342.

Newman, G.R. (1977). Social institutions and the control of deviance: A cross-national opinion survey. European Journal of Social Psychology, 7(1), 39-59.

MacArthur Foundation Research Network on Adolescent Development and Juvenile Justice. (2006). Issue brief #3: Less guilty by reason of adolescence. Retrieved from http://www.adjj.org/downloads/6093issue_brief_3.pdf

Sansone, R.A., & Sansone, L.A. (2011). Gender patterns in borderline personality disorder. Innovations in Clinical Neuroscience, 8(5), 16–20.

Urpelainen, J. (2011). The origins of social institutions. Journal of Theoretical Poli-tics, 23(2), 215-240. doi: 10.1177/0951629811400473..

Misdiagnosing Conduct Disorder
in Hyper-Masculine Environments

Daniel Guzman

The DSM as a Diagnostic Tool
and Cultural Purveyor

The *Diagnostic and Statistical Manual of Mental Disorders (DSM-V)* is a compilation of mental health disorders as standardized by mental health professionals within the United States. The first edition of the DSM was developed in 1952 as a response to U.S. psychiatrists tending to exacerbate mental health disparities post-World War II. Since then, five new editions of the DSM have emerged and been modified to keep up with innovative research and changes in society.

While the DSM has received accolades for creating a compressed, simplified handbook of postulated mental disorders, it has also generated great controversy. Critics have questioned the validity of many of the diagnoses and have argued that there is a lack of scientific and empirical evidence to substantiate its widening claims. The DSM has also been subjected to critique for its blind spots and shortcomings in regards to cultural considerations.

Research has shown that misdiagnosing patients in underrepresented and marginalized communities is a common mistake by clinicians adhering to DSM standards (Lonner & Ibarhim, 2002). Additional studies support this notion and have indicated that the prevalence of misdiagnosing patients in the U.S. has been found to be as high as 47% (Anderson, Hill, & Key, 1989). Due to the alarming percentages of inaccurate diagnoses and the recognized limitations of the DSM, mental health practitioners have every right to question if the manual does more harm than good.

Despite the DSM's controversial reputation, it is still the most widely accepted classification system of diagnoses by psychiatric experts in the United States. Thus, without question, the DSM is a necessary tool and vital component to professionals involved in the caretaking of clients who suffer from mental health disparities. The DSM is an intrinsic feature to the various agents and sectors working for the health care industry. It accessibly establishes a universal lexicon, and its extending influence is corroborated by its popular use in hospitals, clinics, and insurance companies. The importance of having a general text to establish communication practices across several occupations is without question; however rather than accepting the DSM as a hard and fast bible, it should read more like an instruction manual to be tailored on a case-by-case basis to avoid misdiagnosis and ensure best practices.

The collateral effects of misdiagnosing patients warrant personal and societal costs. Individuals who have been misdiagnosed are extremely susceptible to internalizing their disorder and are likely to use their diagnosis as a customary excuse for delayed progression. This adversarial behavior affects the family and poses challenging dilemmas for therapists looking to help clients reach their goals. Moreover, it increases monetary costs and needless spending in both the prison industrial complex and the pharmaceutical industry. In essence, misdiagnosing a patient has serous repercussions on both the individual level and macro level.

The importance of critically analyzing the DSM and the dexterous nature of diagnosing patients is not an easy task for any clinician, but the job bears great significance in providing quality care and reducing the financial implications of misdiagnosis. Given the polemic history of the DSM and its prevalent use to classify individuals who suffer from mental health problems, the nature of this paper will explore the misdiagnosis of African American populations in underserved communities. I will draw from conflict theories to highlight the cultural differences the DSM fails to recognize. Furthermore, I will draw from current research in education and in mass incarceration to relay the symbiotic relationship between conduct disorder and deviant behavior as being a direct response to the economic and societal inequalities found within marginalized communities.

Uneven Playing Field and Cultural Bias
in Social Work Practices

According to a survey in 2003, the National Association of Social Workers (NASW) found that 87% of social work professionals identified as White/Caucasian. Although the study was conducted in 2003, the lack of cultural diversity and ethnic minority in the field of social work highlights the overrepresentation of White social workers in communities of color.

Historically, the field of social work has largely built its theoretical knowledge and concrete practices based on Eurocentric doctrine (Abrams & Moio 2009). The article *Rethinking Work with "Multicultural Populations"* elicits that most medical treatments and diagnosis practices have traditionally experimented with mostly Caucasian populations (Casimir & Morrison, 1993). As a result, the fields of psychology and social work have largely ignored and overlooked cultural issues. Federal studies have analyzed the consequences of this and have advised clinicians to incorporate a more inclusive approach that takes into account the client's social and cultural backgrounds. The nature of operating from a Eurocentric context excludes heterogeneous populations and disregards diverse behaviors, attitudes, and values of clients who are not of European descent.

Both scholars and social workers have challenged social welfare systems on their rigid applications and overemphasis on imposing American middle-class norms on diverse clients and communities of color (Weaver, 1999). In accordance with the traditional underpinnings of social work, Eurocentric biases are a predominant feature in social work practice and have inevitably contributed to deficiency oriented views of individuals and communities of color (Abrams & Moio 2009).

In light of this evidence, we move to assessing the definition and criteria of conduct disorder specified within the DSM-5 handbook. We do this with the intent of juxtaposing the distinct cultural differences found within underserved African American communities and understanding the implications of being a racial minority with a long history of oppression and racism in the United States.

Criteria and Prevalence of Conduct Disorder

According to the DSM, the characteristics of conduct disorder are condensed into 15 specific criteria and divided into four different categories. The categories include: Aggression to People and Animals, Destruction of Property, Deceitfulness or Theft, Serious Violations of Rules. In order for an individual to be diagnosed with conduct disorder three of the 15 criteria must have been violated within a 12-month span and at least one criterion must be present within the past 6 months. Conduct disorder is one of the most prevalent mental health disorders affecting youth and is commonly referred for treatment services. Studies have shown the rates of conduct disorder in desolate urban communities are twice that of suburban communities (Henggeler, Sheidow, 2003).

The DSM briefly mentions that conduct disorder is often misapplied to individuals living in high-crime areas; however, there is a lack of information regarding the societal implications associated with the categorization of conduct disorder. The important point I want to emphasize is that the American Psychiatric Association writes the DSM, and they essentially control and influence clinicians, agencies, and companies whose decisions and polices are involved in the massive mental health industry within the United States. Their hegemonic power to create new labels and define 'normal' behavior should raise several red flags and warrant further investigation of other debatable diagnoses. Thus, this paper will contrast the dominant cultural ideologies highlighted in the DSM with African American culture in underserved communities.

Conduct Disorder and School Suspensions

In contemporary society, young Black males continue to be aggressively stigmatized by unfair stereotypes and institutional discriminatory policies within the education field and the criminal justice system, raising legitimate concerns by the African American community. The public school system has been starkly criticized for disproportionately suspending African American Students, particularly Black males (Townsend, 2000). A study conducted by the U.S. Department of Education Office for Civil Rights in 2014 found that Black students were suspended and expelled at a rate three times greater than white students. Furthermore, the study elicited that 5% of white students are suspended, in comparison to 16% of Black students (Duncan, Lhamon, 2014).

In conjunction with the inconsistency in school suspensions, Black males have been found to receive the diagnosis of conduct disorder at a significantly higher proportion than their Caucasian peers (Clark, 2007). Ogbu (1982) highlights the frustration of African American students needing to familiarize themselves with both their culture and the dominant culture of the school. The cultural discontinuity between African American students and their teachers postulate a feasible reason for the exceedingly high suspension rates. We should note a congruency for the lack of cultural diversity within the education spectrum, as was the case for social workers in their practice, and the extreme shortage of African American teachers in the public school system (Townsend, 2000).

School suspensions increase the probability of falling behind academically and simultaneously intensify the implications and consequences of the school to prison pipeline. It should be noted that self-fulfilling prophecy is a real phenomenon that affects vulnerable populations, and many people who are identified with a particular disorder subsequently internalize the disorder. According to Rios (2009), Black and Latino youth are overrepresented in every major component of the juvenile justice system. The interrelation between school suspensions, mass incarceration, and the negative stigma associated with being diagnosed as having conduct disorder warrants further analysis on the social, environmental, and cultural factors afflicting African American communities.

Conduct Disorder and Code of the Streets Theory

Because of the pervasive societal injustices in historically oppressed communities, which include mass incarceration and racial inequality, an overwhelming majority of Black males living in high incarceration rate communities are forced to adhere to what Professor Anderson has coined the "code of the streets" (2000, p.75). The "code of the streets" exists as a set of norms and unwritten rules within inner city communities that are used to resist dominant mainstream systems and reclaim power in historically disenfranchised communities. Rios (2009) sheds light on the complex masculinity issues governing the "code of the streets": "While wealthy men can prove their masculinity through the ability to earn money and consume products that make them manly, poor young men have to use toughness, violence, and survival as the means of proving their masculinity and resilience." Hence, poor Black males are expected to balance systemic issues of poverty and crime with the social and cultural expectations of mainstream society.

The pressure of adhering to classroom conduct and learning to navigate school spaces for 6 ½ hours a day is overpowered by the fact that impoverished Black males return home to the endemic violence in their neighborhoods. The ubiquitous nature of the "code of the streets" subjugates youth of color to hyper masculine expectations that are in direct conflict with the structures of dominant institutions. It is for these reasons that both social workers and educators must reengineer their attitudes and beliefs regarding what constitutes a conduct disorder based on the context and culture of the specific community. What is emerging in this context is the need to analyze cultural differences between Blacks and Whites to further examine the difficulties social workers and educators face in their methodologies to provide culturally sensitive services, and to abstain from further criminalizing young Black males.

Cultural Differences Between Blacks and Whites

Child-rearing practices of African American parents are vastly different than their White counterparts based on distinctive cultural differences and socioeconomic disadvantages (McNeil, Capage, & Bennett 2002). Numerous studies have noted that African American children are more likely than Caucasian children to live in single parent homes where the mother functions as both the bread winner and disciplinary figure (McBride, Murray, Bynum, Brody, Wilert, & Stephens 2001; Hildreth, Boglin, & Mask 2000). It is crucial to recognize the variables and external factors that lead to dissimilar parenting styles between African American and Caucasian parents.

For instance, African American parents teach their children special skills to cope with living in high-risk environments and thus predictably hold different social values than Caucasian parents based on the historical context of being a racialized minority. Likewise, it is worthy to note that the parallels of growing up without a father in impecunious conditions, regardless of race, places adolescents at a greater risk of developing psychological problems. The absence of a central father figure and the pressure of being a single parent are correlated to increased levels of parental stress. In spite of the social challenges African American mothers face in child rearing practices, African American culture is distinguishably known for people's resiliency in coping with social detriments. The large emphasis on collectivity, rather than individuality, is reflected in the social relationships and kinships established outside of the immediate family. The large support networks include: family members, church members, and neighbors, many of whom serve as caretakers reducing the burden and costs of being a single parent (McNeil et. al., 2002).

Other distinct cultural differences between the Black and White binary include the inclination of African American children to utilize kinesthetic movements in their everyday engagements and learning processes (Hale, 1986). Accordingly, teachers and social workers are susceptible to deeming ordinary behavior that is a distinguishing characteristic of African American culture as inappropriate or deviant based on their subjective experiences and unconscious biases. The socialization and needs of African American children are vastly different than their White peers based on numerous social disparities and cultural differences, as such more exposure to minority cultures and culturally competent training is required to avoid misdiagnosing young Black males in mental health settings and classrooms.

Conclusion

My paper has shed light on the notion that conduct disorder is often attached to young Black males from disenfranchised communities. The cultural discontinuity between those assigning the labels and those internalizing them further stigmatizes them. As I have shown there are significant racial, social, and economic disparities that exist between White and Black families. This is particularly evident in the distinctive child rearing practices of African American mothers and the resilient attitudes inculcated in their children. Moreover, the economic castration, disproportionate suspension rates, and high incarceration rates experienced by African American males are a direct call to action to innovate clinical assessments and educational practices that were traditionally effective in Caucasian populations.

The perilous conflict of navigating hyper masculine environments and conforming to normal student conduct is often a losing battle for young Black males. But for precisely that reason, more responsiveness and advocacy is needed from educators and social workers to bring these issues to the political arena. The mislabeling of conduct disorder and the lack of practitioners challenging dominant ideologies and contributing diverse perspectives to the field of social work suggest changes are necessary to increase culturally sensitive practices and introduce alternative paradigms to working with historically oppressed communities. The features of conduct disorder are one of many responses and reactions to the asphyxiating systemic inequalities inundating African American communities from shaking off the legacy of racism in America.

References

Abrams, L., & Moio, J. (2009). Critical Race Theory and the Cultural Competence Dilemma In Social Work Education. Journal of Social Work Education, 45(2), 245-261.

Anderson, E. (2000). Code of the Street: Decency, Violence, and the Moral Life of the Inner City. New York: W.W Norton.

Anderson, R., Hill, R., & Key, C. (1989). The Sensitivity and Specificity of Clinical Diagnostics During Five Decades: Toward an Understanding of Necessary Fallibility. JAMA: The Journal of the American Medical Association, 261, 1610-1617.

Casimir, G., & Morrison, B. (1993). Rethinking Work with "Multicultural Populations" Community Mental Health Journal, 29(6), 547-559.

Clark, E. (2007). Conduct Disorders in African American Adolescent Males: The Perceptions That Lead to Overdiagnosis and Placement in Special Programs. The Alabama Counseling Association Journal, 33(2), 1-7.

Duncan, A., & Lhamon, C. (2014). Civil Rights Data Collection Data Snapshot: School Discipline. U.S. Department of Education Office for Civil Rights, (1), 1-23. Retrieved April 16, 2015, from https://www2.ed.gov/about/offices/list/ocr/docs/crdc-discipline-snapshot.pdf

Hale, J. (1986). Black children: Their Roots, Culture, and Learning Styles (Rev. Ed., Johns Hopkins Paperbacks Ed.). Baltimore: Johns Hopkins University Press.

Henggeler, S., & Sheidow, A. (2003). Conduct Disorder And Delinquency. Journal of Marital and Family Therapy, 29(4), 505-522.

Hurd, E., Moore, C., & Randy, R. (1995). Quiet success: Parenting Strengths Among African Americans. Families in Society, 76(7), 434-443.

J. Hildreth, G., L. Boglin, M., & Mask, K. (2000). Review of Literature on Resiliency in Black Families: Implications for the 21st Century. 1-6.

Lonner, W. J., & Ibrahim, A. A. (2002). Appraisal and Assessment in Cross-Cultural Counseling. In P. B. Pedersen, J. G. Draguns, W. L. Lonner, & J. E. Trimble (Eds.), Counseling Across Cultures (5th ed., pp. 355–379). Thousand Oaks: Sage.

McBride Murray, V., S. Bynum, M., H. Brody, G., Wilert, A., & Stephens, D. (2001). African American Single Mothers and Children in Context: A Review of Studies on Risk and Resilience. Clinical Child and Family Psychology Review, 4(2), 133-155.

Mcneil, C., Capage, L., & Bennett, G. (2002). Cultural Issues in the Treatment of Young African American Children Diagnosed With Disruptive Behavior Disorders. Journal of Pediatric Psychology, 27(4), 339-350.

Rios, V. (2009). The Consequences of the Criminal Justice Pipeline on Black and Latino Masculinaty. Sage Journal, 623(1), 150-162.

Spencer, M., Lewis, E., & Gutiérrez, L. (2000). Multicultural perspectives on Direct Practice in Social Work. In P. Allen-Meares & C. Garvin (Eds.), The Handbook of Social Work Direct Practice (pp. 131–149). Thousand Oaks, CA: Sage.

Townsend, B. (2000). The Disproportionate Discipline of African American Learners: Reducing School Suspensions and Expulsions. Exceptional Children, 66(3), 381-391.

Weaver, H. (1999). Indigenous People and the Social Work Profession: Defining Culturally Competent Services. Social Work, 44(3), 217-225.

Using a Social Constructionism and Feminist Approach to Criticize the Diagnosis of Reactive Attachment Disorder

Erin K. Murphy

Background of Reactive Attachment Disorder

According to data collected by the U.S. Department of Health and Human Services, the estimated number of children who experience abuse or neglect per 1,000 children dropped from 13 to 12 between 2001 and 2004. However, during the same period, the total number of child fatalities due to abuse rose slightly, from 1,420 to 1,490 in the United States (Tonn, 2006). What these statistics are indicating is that even though the number of children who experience abuse and neglect are fewer, this trauma is more often resulting in death as opposed to a couple of bruises. As disturbing as it is, abuse and neglect towards children are becoming more serious, and it is happening more than reported. Children who have faced severe abuse or neglect are subject to developing psychological disorders that require immediate attention. Severe exposure to trauma or other stressors can explicitly be used as a diagnostic category for several different disorders in the DSM-V, such as Reactive Attachment Disorder (RAD; DSM-V, 2013).

Reactive Attachment Disorder (RAD) was introduced 20 years ago in the DSM-III. It is one of the only disorders that can be applied to infants in addition to being evident before five years old; therefore, diagnosis of RAD in children older than five years old should be used with extreme caution. Only about 10% of severely neglected children are diagnosed with RAD, and it is so uncommon that it is rarely seen in clinical settings (DSM-V, 2013, p. 266). Because of its rarity in clinical settings, the link to proper treatment is not yet fully distinguished.

RAD is one of the most misunderstood and least researched disorders in the entire DSM-V; therefore criteria for RAD are limited. The problem with the diagnostic criteria is that they focus more on a child's deviant social behavior rather than on a child's disturbed attachment behavior, indicating that there is not much research done. There are also problems with the specifications in how the DSM-V identifies "pathogenic care" as a cause for the disorder. Pathogenic care is not well operationalized, and the link to maltreatment is implicit. Moreover, the DSM-V does not specifically define what maltreatment is, or even the longevity or severity of it; all that is said is that severe neglect and abuse are indicators for developing RAD.

The causes of a child being diagnosed with RAD are parents or caregivers not being affectionate, unwanted pregnancy, pre-birth exposure to drugs, alcohol or trauma, neglect or abuse, constant change in daycare or primary caregivers or even moms with chronic depression (DSM-V, 2013). As if their lives weren't hard enough, most of the 10% of children who are diagnosed with RAD are children from the foster care or other institutions and have had multiple caregivers in a shortened time period. The change in caregivers and the maltreatment that most children face in institutions are enabling these children to be more at risk for developing RAD.

This paper will explain the diagnostic criteria of Reactive Attachment Disorder from the DSM-V. It will also discuss findings about Reactive Attachment Disorder and how Social Constructionism and Feminist Theories are relevant to how we understand RAD today. Using Social Constructionism Theory, I will address the issue of how labeling women as the main caregiver can create societal pressures to raise their children perfectly, as well as using Feminist Theory in an attempt to break down the patriarchal barriers of how its oppression towards women affects their bond with their own children.

Reactive Attachment Disorder Diagnostic Criteria

The definition of Reactive Attachment Disorder, according to the DSM-V, is "...characterized by a pattern of markedly disturbed developmentally inappropriate attachment behaviors, in which a child rarely or minimally turns preferentially to an attachment figure for comfort, support, protection, and nurturance" (P. 266). Children diagnosed with RAD had a "consistent pattern of extreme insufficient care, a persistent social and emotional disturbance, and a consistent pattern of inhibited, emotionally with-drawn behavior towards adults or caregivers" (DSM-V, 2013, p. 265).

An important distinction to decipher is that children who were diagnosed with RAD have the capability to form attachments, however, during their childhood they had limited opportunities and limited positive interactions with adults (DSM-V, 2013). In comparison, children who are developmentally incapable of forming attachments should not be diagnosed with RAD. Children, who are diagnosed with RAD lack social functioning, and their capability to form relationships and bonds were never a learned skill set. This distracts the child from having any positive emotions, resulting in negative emotions of fear, sadness or irritability (DSM-V, 2013). Children's diagnosis of RAD usually co-exists with other developmental delays, especially delays in cognition and language (p. 266). What is more important to understand is how these developmental delays and being diagnosed with RAD affect how a child is perceived by society, or more importantly how much is blamed on the child and how much on the mother/caregiver?

Social Constructionism and RAD

Individual behavior is shaped by norms and rules constructed by society. The category that someone is placed in by society shapes his or her behavior and gives meaning and purpose to an individual's life, creating hierarchies. According to Solomon (2008), using an ecological framework is a way in which you can begin to explain human behavior and the "dual perspectives" of different social categories (p.133). Women are concerned with the simultaneous work of caring for their children and work done outside the home. These simultaneous duties interfere with their ability to attach to children in the way that is expected (Solomon, 2008). Due to this, women's supportive needs are being neglected through the price of societal norms and roles that they are required to fulfill and because of this, some children are not getting the required attention they need which puts them at a high risk for RAD.

In the 1950's when attachment theory was developed, working class, people in poverty, or even people of color were not included in the premises of the theory. This causes underprivileged classes to be rendered as inadequate caregivers if they do not reflect the same values as a white-middle class family. However, if attachment theory were derived from people of color, the prevalence or diagnostic criteria of attachment disorders, such as RAD, would look very different. Compared to white- middle class families, work done outside the home is considered a necessary means of financial support and not an option over a stay-at-home parent (Solomon, 2008). These different perspectives according to race and class are socially constructed pressures and can be argued to create the conditions for attachment disorders rather than putting the blame on the mother (Reibstein, 2013).

RAD Through Feminism

The increasing research on attachment disorders while focusing on the historical context of a woman's role in and outside of the home provides insight on where attachment disorders derived from (Solomon, 2008). Due to historical changes in women's labor markets after WWII, the importance of mother-child relationships were criticized (Solomon, 2008). However, women's work in the labor market became an aspect of their family although childcare was still seen as their primary responsibility (Solomon, 2008, p.140). The problem with attachment disorders looked at from a Feminist lens is that it limits our thinking into simplistic terms of predominant attachment styles (Dallos, 2004). Moreover, these simplistic terms are then composited into a mother's mentality that she has to be the best mother.

Due to these pressures, "many mothers experience considerable anxiety of not being a 'good enough' mother...shaped by culturally shared and unrealistic expectations about self-sacrifice, unflinching availability and consistent positive emotions towards their children" (Dallos, 2005, P. 50). The critical blaming model in families causes depression in mothers and may be responsible for transferring attachment disorders to their children (Dallos, 2004). Depressed mothers then find it extremely difficult to describe and acknowledge their own personal emotions that will also affect the relationship with their children. When woman are unable to perform their duties in the home due to stress or lack of support, they are viewed as inadequate mothers, which could lead them into depression. "Depressed mothers have difficulty maintaining interactions with their children...and interferes with parenting and is linked with the development of emotional regulation and behavior problems in children-thus making subsequent parenting even more difficult" (Barth, 2008, p. 98). A feminist perspective faults this view of power imbalances due to patriarchal societies that causes woman to be subdued to psychological problems and not giving their children the attention they need. If woman are subjected to being homemakers, at the same time dealing with their own mental illnesses, then attachment disorders are more likely to be present.

In attachment theory, there is a presumption that children need to be accompanied by power and guidance in order to feel protected. Consequently, women are struggling with their own oppressions that limit their ability to give children that power and guidance they need, therefore putting their own children at risk for developing RAD. Placing ideological labels such as domestic, passive, subordinate or even emotional creates conflicts with the idea that woman are overwhelmingly responsible for the outcome of their children. "Such terms...may become internalized and over time come to shape and eventually consume a person's identity to the point where other aspects of their lives...become marginalized" (Dallos, 2004, P. 41).

Another feminist perspective to consider is that even though the father might not be the primary caregiver, the father's relationship with the mother can still have an influence on the child (Dallos, 2004). According to Solomon (2008), "even when fathers make considerable contributions to child care, it is likely that mothers remain accountable for the quality of that care and their children's overall healthy development" (p. 141). For example, if the father has a hostile relationship with the mother, it may propel the mother to not meet their child's needs; however, if the father is affectionate towards the mother, those qualities will relay into the relationship with the child (Dallos, 2004). We can clearly see, either indirectly or directly, the pressure under which women have to form the attachments with their children.

So What?

Feminist Theory states that there is a basis of subordination of women to men as the Social Constructionist Theory identifies how the woman's subordination is then labeled as a "homemaker." So what do the Feminist and Social Constructionist perspectives mean for diagnosing children with RAD? If society puts less pressure on women to be perfect mothers, does this mean that there would be fewer children being diagnosed with RAD? What would happen if women ignored the stereotypical ideologies and didn't seem as though childcare was their main priority? Would people begin to see them as deviant or as independent women? What would the diagnostic criteria or even ideological labels look like to both men and women if the mother and father both shared the blame for their children being diagnosed with an attachment disorder and would this lead to a shift in power? All of these are great questions to consider in order to begin to understand where our society came from and where we are headed or in what ways we can start being agents for change.

Today more than ever, more dads are becoming stay-at-home dads, which would be unheard of 75 years ago. This shift in household roles will hopefully help society to see how woman have been subjected to ideological labels and notice how woman should not solely be blamed for a diagnosis of RAD. At the pace society is changing, while more dads are becoming stay-at-home dads, will it ever become that dads are subjected to the same labels as woman are today? During the second wave of the feminist movement, some woman took the stance that motherhood was a significant source of woman's oppression (Liss & Erchull, 2012). Could this ever be applied to fathers in the same fashion if the new trend is dads staying at home if this becomes their primary responsibility? Due to the fact that the United States lives in a patriarchal society, that is not likely to happen. That would take centuries to completely change into a matriarchal society, however, a shift in perspectives is a much more obtainable request. We could begin to see a shift from women being the sole caretakers to having shared responsibilities.

When attachment theory was first introduced, it focused on the mother-child dyad, however, now we have seen a shift in changes where a child can form a bond with any adult. With more and more dads becoming caregivers, according to Rudolph Schaffer, as cited in Solomon 2008, there is "nothing to indicate any biological need for an exclusive primary bond [and] nothing to suggest that mothering cannot be shared by several people" (P. 139). In order to accept attachment theory, we have to look at socio-political power relations and how they define the limits of normality (Solomon, 2008). This leads to questioning the power relationships within the family and the immediate needs of children. Children do not need an attachment primarily from the mother, rather an attachment from any adult. As long as the children are accompanied by supervision, love and guidance, they will learn how to have relationships. This theory clearly indicates that the relationship solely of the mother is not needed. Moreover, Pocock 2010, describes this concept as "attempt[ing] to find the best fit [for] each other" (P. 304). This is an important concept to consider because it is a common myth that children need warmth and compassion from their mothers. This is a direct indication to when children are diagnosed with attachment disorders, people automatically shift their attention to the mothers as if they did something wrong.

Concluding Thoughts

Reactive Attachment Disorder is one of the most misunderstood diagnoses in the entire DSM-V because it is so rare. Attachment disorders are so important to understand because these children are our future. With the increasing number of fatalities of abused children, we have to pay attention to the bigger picture of how our societal views have obstructed the view of what's really important. We should be focusing on how the language we use oppresses women and creates this never-ending boundary to conform to becoming homemakers. No matter if you are a male or female, having dual responsibilities is never easy, and when you add pressure to conform, it creates an inner struggle with the mothers; even leading some into depression. How is a mother supposed to form an attachment with her children while she can't even take care of herself? Mothers should not be deemed the sole responsibility of caretaking. As research develops, they conclude that a child can learn to form attachments and bonds from anyone who gives them the time of day to support them.

Bottom line, Social constructionist and Feminist influences have gone too far (Pocock, 2010). People's perspectives can change the way some people are oppressed and others privileged (Solomon, 2008). Society has to change their own views from an unequal distribution of labor and oppressive language to a shared responsibility and equality in the home. This will ultimately trickle down the belief system of what a woman and man's roles are. Through interactions, people define who they should be and how they should act. With an understanding of how norms affect people, we could start a promotion for change. The necessity of change begins with drawing "attention to the socio-political, economic and historic conditions that produces policy and practice... and is shaped and re-shaped by ideas and conditions over time" (Solomon, 2008,p. 135). In other words, attachment disorders, through the lenses of a Social Constructionism and Feminism approach may be interpreted as a privilege claim relating to the unequal distribution of historical power and provide reasons why some children are more likely to become child welfare clients or children with RAD and others not (Solomon, 2008, p. 138). It gives us a perspective on how distribution of power not only oppresses some and privileges others but also how it could lead more children into attachment disorders. Be an agent for change, stop the labels and start worrying about what can be done to help these children rather than pointing fingers at one another.

References

American Psychiatric Association. (2013). *Diagnostic and statistical manual of mental disorders* (5th ed.). Washington, DC: American Psychiatric Association.

Barth, R. P. (December 07, 2009). Preventing Child Abuse and Neglect with Parent Training: Evidence and Opportunities. Future of Children, 19, 2, 95-118.

Dallos, R. (February 01, 2004). Attachment narrative therapy: integrating ideas from narrative and attachment theory in systemic family therapy with eating disorders. Journal of Family Therapy, 26, 1, 40-65.

Hosking, D. M., & Morley, I. E. (September 01, 2004). Social constructionism in community and applied social psychology. Journal of Community & Applied Social Psychology, 14, 5.)

Liss, M., & Erchull, M. J. (August 01, 2012). Feminism and Attachment Parenting: Attitudes, Stereotypes, and Misperceptions. *Sex Roles : a Journal of Research, 67,* 131-142.

Pocock, D. (January 01, 2010). The DMM - wow! But how to safely handle its potential strength? Clinical Child Psychology and Psychiatry, 15, 3, 303-11.

Reibstein, J. (November 01, 2013). Commentary: A different lens for working with affairs: Using social constructionist and attachment theory. Journal of Family Therapy, 35, 4, 368-380.

Solomon, B. P. D. M. S. W. (January 01, 2002). A Social Constructionist Approach to Theorizing Child Welfare. Journal of Teaching in Social Work, 22, 131-149.

Tonn, J. L. (2006). Child Abuse and Neglect. *Education Week, 25*(33), 19.

PTSD: The Chicken or the Egg?

Jack Knight

What came first, the chicken or the egg? It is an age old question which lends itself to much debate. However, when the question is assigned to Post Traumatic Stress Disorder (PTSD), the conversation looks a little different. PTSD is caused by a traumatic event or stressor which, in the chicken and egg scenario, would it be considered the chicken or the egg? How about the severity of the stressor? Or the lack of support system? What is it that actually plays into the development of PTSD? The point of this paper is to show the varying issues that play into the symptomology of PTSD, and thus, the diagnosis of PTSD.

There is a reality to PTSD that should be taken into consideration. When a traumatic event or stressor is dealt to an unsuspecting individual, in which they then respond with enough impaired functioning to meet the criteria for a diagnosis of PTSD, then there should be something in place to explain the situation better, correct? Is that not what the criteria attempt to do? It is reasonable to have criteria for PTSD, is it not? But what this paper questions, among other questions, is can we understand the full issue by only hearing the account following a traumatic event? It is, after all, POST traumatic stress disorder. And how can we be so sure that symptoms were not caused because of vulnerabilities before the traumatic event? Those questions will be addressed soon enough. It might be helpful to encounter the history of the diagnosis before we ask the hard hitting questions.

Posttraumatic stress disorder was introduced in 1980 in the third edition of the American Psychological Association (APA) *Diagnostic and Statistical Manual of Mental Disorders* (DSM-III). This was primarily in response to many veterans of the Vietnam War presenting with symptoms that were not fulfilling criteria in the DSM-II (Elwood, et al., 2008; Bloom, 2000). As the new diagnosis became understood for veterans, many victims of other traumatic stressors were viewed as presenting similar symptoms, such as victims of assault, rape, or natural disasters. The initial diagnostic criteria stated that an individual must have experienced a stressor that would evoke significant symptoms of distress in almost everyone, followed by the experience of three symptom clusters: re-experiencing, numbing, and miscellaneous symptoms (Elwood et al 2008; APA 1980). The DSM IV revised it to include re-experiencing (flashbacks or nightmares), avoidance and numbing (restriction of affect, avoidance of traumatic cues), and increased arousal symptoms (exaggerated startle response, difficulty sleeping), but kept the symptom clusters as previously introduced (Elwood et al 2008; APA 1994). However, a key difference in the DSM-IV was that it modified the focus from being on the response of the person to the subjective experience of the event, claiming that in order to fit the criteria, an individual must have experienced, witnessed, or have been confronted by an event that involved actual or threatened danger and evoked feelings of helplessness, fear, or horror (Elwood et al 2008; APA 1994).

Most recently in the DSM V, published in 2013, according to Jones and Cureton (2014) the foremost change with the diagnosis of PTSD was its assignment to an "innovative diagnostic category" called Trauma and Stressor-Related Disorders and Dissociative Disorders (Jones & Cureton. 2014). Removing it from the Anxiety Disorders category of the previous DSM editions proved to be significant, perhaps innovative, because as they state, "this highlights the central importance of the predisposing stressor," and furthermore, this shift recognizes that an event or stressor can be influential for an entire class of conditions that might threaten a person's functioning (Jones & Cureton, 2014). The argument in the changes to the DSM V is that it makes it more subjective and more comprehensive in the explanation of what a stressor might be. According to the authors, citing Mcnally (2009), the broad definition of PTSD in the DSM-IV led to an overdiagnosis of PTSD because the loose definition could include less threatening events (Jones, Cureton. 2014; Mcnally, 2009).

Why is this important? For one, since the beginning of its creation as a diagnosis, there has been a lot of fluctuation in understanding PTSD. Anxiety is seen to be connected to PTSD, but it wasn't the whole story. Perhaps PTSD is more about a stressor and how that affects the subjective experience, so let's create a whole category to explain it. And two, is it the severity of the trauma, or is it the mere presence of a stressor that is important with diagnosis of PTSD? What's the harm, anyways, in diagnosing somebody who develops symptoms of PTSD after a significant stressor?

There are practical reasons to diagnose PTSD. After all, our model of service is managed care--Clinicians have to make a living. If somebody fits the criteria of PTSD, why not diagnose him or her with PTSD? In Illinois, the adult Medicaid eligibility for a diagnosis of PTSD is in the same eligibility category for adjustment disorder, which is a common alternative to many diagnostic decisions (Illinois Department of Human Services, retrieved April 12, 2015). In other words, someone with a diagnosis of adjustment disorder would obtain the same benefits as the treatment of someone with PTSD. The case where this has caused significant issues, however, is in Veterans Affairs.

According to the 2014 Congressional Budget report *Veteran's Disability Compensation: Trends and Policy Options*, compensation to veterans is determined on a scale of zero to 100, with higher ratings reflecting greater severity of disability (Congressional Budget Office, 2014, p 7). In the case of veterans, it appears that diagnosing PTSD for those who present symptomology is reasonable. However, for Mental Health compensation, determination of disability is based on occupation and social impairment, and additionally, the report states that mental health standards are held to stricter standards than those of physical impairments (Congressional Budget Office, 2014, p 7). Basically, the worse off a veteran's presentation of functioning, the more they might qualify for disability. However, the qualifications are held under larger scrutiny than those for physical disability. People can trust seeing a physical impairment and determine disability easier than they can trust the self-disclosed impairments of mental health.

A 2014 *Washington Post* article entitled, "As Disability Awards Grow, So Do Concerns over the Veracity of Veterans' PTSD Claims," quotes clinicians of VA hospitals across the country challenging the increase of PTSD diagnosis in veterans. This was not to discuss the significance in real symptomology, but rather as the author reports, symptomology is exaggerated by the veteran's account. A psychologist at the University of Hawaii who worked with veterans for 15 years in the VA system was quoted to say about this exaggeration, "a large chunk of these patients are flat-out malingering." The figures in this article report that disability awards for PTSD have grown nearly fivefold in the past 13 years, totaling over 656,000 veterans on disability rolls compared to 133,745 in 2001 (Zarembo, 2014, October 14). Remember, PTSD has only been a diagnosis since 1980. This means the 133,745 in 2001 consists of veterans from every conflict before 2001.

In 2008 a staff psychologist at the Central Texas VA Health Care System advised staff to diagnose adjustment disorder rather than PTSD. Although this email communication was only meant for interoffice purposes, it created a nation-wide media frenzy as it was initially leaked (Department of Veterans Affairs, 2009). Those opposing the notion of diagnosing adjustment disorder rather than PTSD accused the VA of the intention to limit benefits that are due to veterans suffering from PTSD. Supporters of this notion professed a caution against becoming too quick in diagnosing PTSD, which would run the risk of treating veterans as if they had PTSD, when in reality, they did not.

So, there is a practicality to diagnosing PTSD, but as this information presents, and as the staff psychologist in Texas intended to present, PTSD is a complex diagnosis. But why not be liberal in diagnosing veterans with PTSD? Even if they do fabricate their story to fit the mold of the symptomology, can we give them the benefit of the doubt that they experienced something traumatic as a soldier? That by being a part of a system that is solely built on the destruction, killing, and overtaking, of political enemies, it is not enough to make a person go "insane," for lack of a better word. Some people would welcome a diagnosis of PTSD. By defining some of the feelings they might be feeling, they might become appreciative of a diagnosis. A person could feel that it is not just he or she who feels that way, but enough others to justify diagnosis criteria in a reputable diagnosing tool, such as the DSM. They might search out support groups or other forms of treatment on account of this information. They now have a definition for the ambiguous feelings, which is positive.

However, there is a risk of stigma for a patient with the diagnosis of PTSD. This risk lies in the pathologizing of victimization. Basically, it is one thing to develop symptoms after being the victim of a traumatic event, but to then become labeled as having a mental illness becomes another thing in its own right. And although dysthymia and adjustment disorder are disorders in the DSM, it can be argued that such diagnoses do not hold the weight and complexity of PTSD. Furthermore, who doesn't have dysthymia as a human being at some point in their lives? The focus of this paper will now discuss the pre-trauma vulnerabilities, which is as important as the Post-trauma discussion, and may be the etiology of post-trauma symptomology.

It is acknowledged that most people who experience a traumatic event or stressor will not have the symptom presentation of PTSD. In fact, it is reported that 74% of women and 81% of men will experience a traumatic event in their lifetimes (Bomyea, Risbrough, Lang, 2012; Kessler et al., 2005; Stein, Walker, & Hazen, 1997). However, only a "relative minority" of trauma-exposed individuals actually develop PTSD (Bomyea, Risbrough, Lang, 2012; Kessler, Sonnega, Bromet, Hughes, & Nelson, 1995; McNally, Bryant, & Ehlers, 2003). How is this explained?

There is the notion that symptomology of PTSD is created by the severity of the traumatic event. And although there is a reasonable understanding for this argument, is severity as important as the notion presents? It is true, there are worse things to go through than say, a tornado, but would surviving a tornado bring a sense of shame that sexual assault might bring? And what about blame? Hardly any blame comes from experiencing a natural disaster. But surviving a natural disaster fits the criteria of a traumatic event in the DSM V; and rightly so. Surviving a natural disaster can instill chronic fear and anxiety; symptomology of PTSD. The key to this argument is that PTSD is a subjective diagnosis, which the DSM tries to address but can only go so far in the attempt to describe a person's experience. Diagnosing PTSD is not about debating if a person experienced a severe enough trauma. It is, in the end, about defining the best treatment. But how to explain why some people develop PTSD while others do not? Is resilience a process after the trauma, or is resilience part of the faculties already instilled in the individual before the trauma?

Resilience is defined by Mancini and Bonanno (2006) as the maintenance of "relatively stable, healthy levels of psychological and physical functioning" along with, "the capacity for generative experiences and positive emotions" (Mancini, Bonanno, 2006; Bononno, 2004). Again, this definition is meant to explain resilience post trauma, but this does not sound like the "relative minority" that are experiencing symptomology. This definition appears to describe the experience of a "healthy" individual, and according to Gabbard (2014), "incidents of PTSD are actually rather low among people who are healthy before experiencing the trauma" (Gabbard, 2014, p. 283). So what are the vulnerabilities that would impact the development of PTSD if a person were not "healthy" before a trauma? Gabbard goes on to list individual meanings, genetics, and environmental factors in one's history, as factors that might cause symptoms (Gabbard, 2014).

Early theories proposed that PTSD symptoms were caused when preconceived beliefs about the world became shattered in the event of a trauma. In other words, a person's outlook on the world might be positive until a traumatic event disrupted the schema and proved otherwise, and as a result of the shock the person would then develop symptoms of PTSD (Elwood et al, 2008; Epstein, 1991; Janoff-Bulman; 1992, McCann & Pearlman, 1990). While this might be an early attempt to hypothesize symptomology, it did inspire others to discuss cognitive vulnerabilities more compellingly. Elwood et al. (2008) define cognitive vulnerabilities as "characteristic possession of biased beliefs or cognitive patterns," and furthermore, cognitive vulnerabilities then mediate the relation between environmental events and emotional responses (Elwood et al, 2008; Riskind & Alloy, 2006). In other words, they are describing the formulation of habitual pre-trauma cognitive patterns. They go on to make an important distinction, "Cognitive vulnerabilities are assumed to remain latent until activated by sufficient stress or negative life events" (Elwood et al 2008). These biased beliefs and patterns of behavior are lying dormant until aroused by an intrusive event, causing the beliefs to rise to the surface. And furthermore, the emotional response will be mediated by these vulnerabilities. This makes the diagnosis of PTSD that much more complex and should be considered while a patient is describing symptoms.

Pre-trauma vulnerabilities are also biological and hereditary. Sherin and Nemeroff (2011), in their attempts to examine neurological vulnerabilities in symptom development, confirm the notion that pathology lays dormant until activated by a stressor stating, "Certain abnormalities in the patient with PTSD simply represent pre-existing pathology that is functionally dormant until released by trauma exposure and detected thereafter upon investigation" (Sherin & Nemeroff, 2011). In the development of PTSD, certain neurological abnormalities are identified to increase such a risk. The authors include, among other pre-existing factors, low cortisol levels, smaller than average hippocampal volume sizes, and exaggerated amygdala activity. The cortical, hippocampus, and amygdala regions are part of a neural network interconnected to mediate adaption to stress and fear conditioning, among other functions. Changes in these circuits have been proposed to have a direct link to the development of PTSD, report the authors. These neurological vulnerabilities are not only biological, but possibly, components such as the smaller volume of the hippocam-hippocampus, may also be linked to hereditary factors. Although the research on heredity and the predisposition for developing PTSD is still a work in progress and, brings much debate, the authors report that family and twin studies have long suggested a heritable contribution to the development of PTSD (Sherin & Nemeroff, 2011).

Now, resilience can be a process post-trauma, but is the minimal presence of pre-trauma vulnerabilities also part of resiliency? If a person has a secure attachment in pre-trauma, can this be considered a form of resiliency? That is not to say that having an insecure attachment is the cause of developing mental illness, but it is part of what amplifies the effects of attachment experiences on the way to psychopathology (Mikulincer & Shaver, 2012). In a 2009 interview with the blog *Neuronarrative*, Dr. Daniel Sonkin reports that 60-65% of the population have a secure attachment (Neuronarrative, 12, January, 2009). Other research studies suggest a lower account. While this reputable study revolved around attachment theory and relationships, Hazen and Saver (1987) reported secure attachment at 56% of the population (Hazen & Shaver, 1987). Whether the rates in reality are on the low or the high end of the spectrum does not negate the value of secure attachment when discussing the development of PTSD symptomology.

Louis Cozolino (2010), in his book *The Neuroscience of Psychotherapy*, states that a "healthy functioning requires proper development and functioning of neural networks that organize conscious awareness, behavior, emotion, and sensation" (Cozolino, 2010, p.20). This healthy functioning is brought on by a secure attachment, which is created by the responses of "good enough" parenting in the care of the infant. As Cozolino reports, this sets up the neural networks to expect positive response and outcomes, and becomes the "sensory-motion-emotional" background of our experience (Cozolino, 2010, p.23). This becomes the habitual nature of experience, which then creates agency because there is an expectation that the stressor will be manageable, and thereby, symptomology of PTSD would not be likely due to the minimal presence of vulnerabilities before the stressor. This is perhaps the reason why the majority of the population go through a traumatic event but do not develop symptoms of PTSD.

Cozolino goes on to explain that without a secure attachment process, the brain organizes defense mechanisms in order to reduce anxiety. These defenses become circuited in our unconscious memory, which controls anxiety and fear (Cozolino, 2010; Critchley et al, 2000). And these defenses in our neural network become the framework of our behaviors, such as what we choose to approach or avoid, what our attention is drawn to, and the assumptions we use to organize our experiences. He goes on to say, "These neural and psychic structures can lead to either psychological and physical health, or illness and disability" (Cozolino, 2010, p.23). In other words, with a secure attachment comes the formation of positive expectation and agency, and without it comes defense coping strategies that become vulnerabilities in the face of trauma which, if an event should happen, would not set up adequate measures of coping with the stress caused by such an event.

The reality is that the diagnosis of PTSD fits a functional purpose. Clinicians, mental health organizations, insurance companies, and veterans, all fit into the managed care system. For veterans, there is considerably more interest in obtaining a diagnosis of PTSD compared to that of adjustment disorder. But for a victim of abuse, a diagnosis of adjustment disorder might be more beneficial with regards to pathologizing victimization. The history of PTSD shows that it is not only early in its course, but is also complicated by the subjective experience; not to mention the reliance on the patient's account. However, even if post-trauma symptomology is present, it is the pre-trauma vulnerabilities that made it so. Perhaps it matters more who a person was pre-trauma rather than POST trauma, and would this negate the diagnosis of PTSD?

In conclusion, perhaps the egg is defined by the pre-trauma vulnerabilities and the chicken is symptomology. But doesn't the DSM-III initially give language to the diagnosis, so would that be considered the egg? In that case, would the DSM-III have language to give if it weren't for the Vietnam veterans who initially presented the symptoms? Would this then be the egg? And there it is, just as in the typical chicken and egg question, one finds oneself in a circular conundrum. The point here is that PTSD is complex and confusing for all involved. The real question is: are clinicians willing to risk stigma in the diagnosis of this mental illness? PTSD is both chicken and egg, both subjective and objective, both pre and post, both individual and systematic. How easy is it then to diagnose PTSD? Probably as easy as solving the chicken or the egg question.

References

American Psychiatric Association (1980). Diagnostic and statistical manual of mental disorders, 3rd Ed. Washington, DC: American Psychiatric Association.

American Psychiatric Association (1994). Diagnostic and statistical manual of mental disorders, 4th Ed. Washington, DC: American Psychiatric Association.

Bloom, S. L. (2000). Our hearts and our hopes are turned to peace: Origins of the International Society for Traumatic Stress Studies. In A. Y. Shalev, R. Yehuda, & A. C. McFarlane (Eds.), International handbook of human response to trauma (pp. 27□ 50). Dordrecht, Netherlands: Kluwer Academic Publishers

Bomyea, J. Risbrough, V. Lang, A. (2012). A consideration of select pre-trauma factors as key vulnerabilities in PTSD. *Clinical Psychology Review*, 32(7), 630-641.

Bonanno, G.A. (2004). Loss, trauma, and human resilience: Have we underestimated the human capacity to thrive after extremely aversive events. American Psychologist, 59, 20–28.

Congressional Budget Office (2014). *Veterans' Disability Compensation: Trends and Policy Options*. (Pub. No. 4617) Retrieved April 17, 2015, from, https://www.cbo.gov/sites/default/files/45615-VADisability_2.pdf

Cozolino, L. (2010). *The Neuroscience of Psychotherapy: Healing the Social Brain*. (2nd ed.) New York, NY: W.W. Norton and Company

Department of Veterans Affairs Office of Inspector General (2009). *Healthcare Inspection: Allegations of Mental Health Diagnosis Irregularities at the Olin E. Teague VA Medical Center Temple, Texas*. (Report No. 08-02089-59) Retrieved April 17, 2015, from, http://www.va.gov/oig/54/reports/VAOIG-08-02089-59.pdf

Elwood, L. Hahn, K. Olatunji, B. Williams, N., (2009). Cognitive vulnerabilities to thedevelopment of PTSD: A review of four vulnerabilities and the proposal of an integrative vulnerability model. *Clinical Psychology Review*, 29 (1), 87-100.

Epstein, S. (1991). Impulse control and self-destructive behavior. In L. P. Lipsitt, & L. L. Mitnick (Eds.), Self-regulatory behavior and risk takings: Causes and consequences (pp. 273□ 284). Norwood, NJ: Ablex.

Gabbard, G., (2014). *Psychodynamic Psychiatry in Clinical Practice.* (5th ed.) Washington DC: APA Publishing

Illinois Department of Human Services. Consumer eligibility, enrollment and benefit status. Retrieved April 17, 2015, from Illinois Department of Human Services website: https://www.dhs.state.il.us/page.aspx?item=51784

Janoff-Bulman, R. (1992). Shattered assumptions. New York: The Free Press

Jones, L. Cureton, J (2014). Trauma redefined in the DSM-5: rationale and implications for counseling practice. *The Professional Counselor,* 4(3), 257-271.

Kessler, R. C., Chiu, W. T., Demler, O., & Walters, E. E. (2005). Prevalence, severity, and comorbidity of twelve-month DSM-IV disorders in the National Comorbidity Survey Replication (NCS-R). Archives of General Psychiatry, 62, 617–627.

Kessler, R. C., Sonnega, A., Bromet, E., Hughes, M., & Nelson, C. B. (1995). Post-traumatic stress disorder in the National Comorbidity Survey. Archives of General Psychiatry, 52, 1048–1060.

Mancini, A. Bononno, G. (2006). Resilience in the face of potential trauma: clinical practices and illustrations. *Journal of Clinical Psychology.* 62(8) 971-985

McCann, I., & Pearlman, L. A. (1990). Psychological trauma and the adult survivor: Theory, therapy, and transformation. New York: Brunner/Mazel.

McNally, R. J. (2009). Can we fix PTSD in DSM-V? Depression and Anxiety, 26, 597–600. doi:10.1002/da.20586

McNally, N. J., Bryant, R. A., & Ehlers, A. (2003). Does early psychological intervention promote recovery from post-traumatic stress? Psychological Science in the Public Interest, 4, 45–79

Mikulincer, M. Shaver, P. (2012). An attachment perspective on psychopathology. *World Psychiatry,* 11(1), 11-15

Neruonarrative, (2009, January 12). Attachment theory and the brain:an interview with Dr. Daniel Sonkin. *Neronarrative blog.* Retrieved April 17, 2015, from https://neuronarrative.wordpress.com/2009/01/12/attachment-theory-and-the-brain-an-interview-with-dr-daniel-sonkin/

Riskind, J. H., & Alloy, L. B. (2006). Cognitive vulnerability to psychological disorders: Overview of theory, design, and methods. Journal of Social and Clinical Psychology, Special issue: Cognitive vulnerability to psychological disorders, 25, 705□ 725

Sherin, J. Nemeroff, C. (2011). Post-traumatic stress disorder: the neurobiological impact of psychological trauma. *Dialogues in Clinical Neuroscience*, 12(3), 263-278.

Stein, M. B., Walker, J. R., & Hazen, A. L. (1997). Full and partial posttraumatic stress disorder: Findings from a community survey. American Journal of Psychiatry, 154, 1114–1119

Zarembo, A. (2014, October 16). As disability awards grow, so do concerns over the veracity of veterans' PTSD claims. *Washington Post*. Retrieved on April 17, 2015, from, http://www.washingtonpost.com/national/as-disability-awards-grow-so-do-concerns-over-the-veracity-of-veterans-ptsd-claims/2014/10/16/8bf577bc-1e64-11e4-ab7b-696c295ddfd1_story.html

Avoidant Personality Disorder:
A Socially Constructed Stigma

Jessica Snype

Mental health is an often misunderstood and misappropriated term that seeks to describe an individual's level of psychological wellbeing. Since the onset of deinstitutionalization in the 1950s, the mental health field has been under much scrutiny as it attempts to understand and treat individuals with mental disorders. As such, it has become necessary for mental health professionals to rely on clinical diagnoses in order to be regarded as legitimate. This has led to an increase in the number of individuals diagnosed with a mental disorder. It is reported that the number of individuals who have been diagnosed with a debilitating mental illness, qualifying them for SSI or SSDI, has increased more than twofold between 1987 and 2007 (Angell, 2011). The U.S. Department of Health and Human Services' 2012 National Survey on Drug Use and Health shows that 18.6% of American adults had experienced any mental illness during the year (2013). Statistics like these have caused critics of the mental health field to accuse mental health professionals of over-diagnosing mental illnesses, often linking these diagnoses with an increase in reliance on prescription medication.

At the heart of many of these critiques is the *Diagnostic and Statistical Manual of Mental Disorders* (*DSM*). The *DSM* is a guide to mental disorders that lists "associated criteria designed to facilitate more reliable diagnoses of these disorders" (American Psychiatric Association [APA], 2013b, p. xli). Many believe that this manual has made it too easy for doctors to diagnose patients and has lead to an epidemic (Frances, 2010; Gray, 2013; Mientka, 2013). While diagnoses of major depressive disorder and generalized anxiety disorder are the most prevalent and draw the most criticism, personality disorders are among the most misunderstood and stigmatized diagnoses. In a 1988 study conducted by Glyn Lewis and Louis Appleby, psychiatrists regard individuals with a diagnosis of a personality disorder as "more difficult and less deserving of care...manipulative, attention-seeking, annoying, and in control of their suicidal urges and debts" (p. 44). This blend of over-diagnosis and pejorative judgment causes individuals diagnosed with a personality disorder (over 9% of the U.S. adult population) to have difficulty seeking treatment and successfully managing their symptoms (National Institute of Mental Health, 2007).

Avoidant personality disorder has been diagnosed in 2.4% of the U.S. adult population (APA, 2013b, p. 674). It is characterized in the *DSM-V* by "a pervasive pattern of social inhibition, feelings of inadequacy, and hypersensitivity to negative evaluation" (APA, 2013b, p. 672). In order to be diagnosed with the disorder, individuals must exhibit four (or more) of the APA's seven distinguishing characteristics: avoiding occupational activities, unwillingness to get involved with others, resisting intimate relationships, preoccupation with criticism, inhibition in new social settings and activities, and viewing oneself as socially inept (p. 672-673). Individuals diagnosed with avoidant personality disorder are overly restrained in social situations due to anxiety and fear of criticism or rejection. This leads to limited and often strained relationships, as the individual is hesitant to seek social events. An anecdotal description of the disorder describes these individuals as having "a very fragile ego, self-image, or understanding of how relationships are to operate. Many are loners or isolators who are too fearful to enter relationships or maintain the one's [*sic*] they already have" (Hill, 2014).

A diagnosis of avoidant personality disorder is not usually made until early adulthood, when it is clear that the symptoms are persistent and not simply characteristic of a developmental stage (APA, 2013b). As children, individuals with avoidant personality disorder are very shy, but, unlike other children, they do not grow out of their shyness (Long, 2015). As adolescents, they may initially be diagnosed with social anxiety disorder, but as it increases with age, a more acute diagnosis of avoidant personality disorder can be made. Often, as individuals with avoidant personality disorder age, they are able to learn coping mechanisms to function relatively highly in day-to-day life. Symptoms are less detectable in older adults, as they have become successful at managing or masking them; however, they are still present. When considering the diagnostic criteria for avoidant personality disorder, it is necessary to recall the *DSM-V*'s description of a personality disorder: "an enduring pattern of inner experience and behavior that deviates markedly from the expectations of the individual's culture, is pervasive and inflexible, has an onset in adolescence or early adulthood, is stable over time, and leads to distress or impairment" (APA, 2013b, p.

645). As such, it is important to differentiate between avoidant personality disorder traits and those that are caused by other disorders or life events. For example, an individual experiencing a major depressive episode may withdraw from social situations and feel worthless, but this is a relatively short-term experience and is not symptomatic of a personality disorder.

A study published in 2005 by Andrew E. Skodol, et al., called the Collaborative Longitudinal Personality Disorders Study (CLPS), analyzed the nature, course and impact of personality disorders. In general, the CLPS reports that personality disorders consist of "maladaptive trait constellations that are stable in their structure (individual differences), but can change in severity or expression over time" (Skodol, et al., 2005, p. 493). Essentially, there are long-term, persistent exaggerated characteristics that cause a range of behaviors "that are attempts at adapting to, defending against, coping with, or compensating for these pathological traits" (p. 493). In individuals with avoidant personality disorder, the most stable criterion was feeling socially inept and inadequate which, in turn, caused avoiding jobs with interpersonal contact (a behavior, not a characteristic) to be the least stable criterion (p. 493).

The persistence of symptoms in an avoidant personality disorder diagnosis is perhaps one of the only differentiating characteristics between avoidant personality disorder and social phobia. Personality disorders, as previously mentioned, differ from other mental disorders because they are "pervasive and inflexible," lasting throughout the duration of an individual's life (APA, 2013b, p. 645). Dr. Phillip W. Long describes avoidant personality disorder as "a persistent, generalized form of social anxiety disorder (social phobia)" (2015). The two disorders have such overlapping symptoms that the *DSM-V* (2013b) even speculates that "they may be alternative conceptualizations of the same or similar conditions" (p. 674).

The *DSM-V* (2013b) diagnostic criteria for social phobia include marked fear regarding social situations, anxiety about being negatively evaluated, fear and anxiety during or avoidance of social situations, persistence over six or more months, and clinically significant distress or impairment (APA, p. 202-203). These criteria are very similar to those necessary for a diagnosis of avoidant personality disorder, and there are very high comorbidity rates with avoidant personality disorder and social phobia. There are estimates that 22 to

89% of individuals diagnosed with social phobia are also diagnosable with avoidant personality disorder (Tillfors & Ekselius, 2009, p.27).

Though more research on the topic must be done, there are many who argue that social phobia and avoidant personality disorder are manifestations of the same disorder. In 1986, Samuel M. Turner, Deborah C. Beidel, Constance V. Dancu, and Dana J. Keys published a comparative study of the pathology of social phobia and avoidant personality disorder. Not only did they criticize the *DSM-III-R*, an earlier version of the *DSM-V*, for implying that social phobia is not a pervasive condition (p. 393), but they noted "no differences between the two disorders" in terms of psychophysical and cognitive variables (p. 394). They did report behavioral differences and differences in self-reported inventories. These differences do not necessarily mean that there are two different disorders, but that there are (at least) two manifestations of what is potentially the same disorder.

Samuel M. Turner, Deborah C. Beidel, and Ruth M. Townsley (1992) conducted an experiment to compare social phobia and Avoidant personality disorder after the *DSM-III* was re-released and revised. They concluded that "social phobia and APD differ primarily in the severity of social anxiety and social functioning" (p. 331). Elissa J. Brown, Richard G. Heimberg, and Harlan R. Juster (1995), using the same edition of the *DSM* and the corresponding criteria, reported statistically insignificant difference in functioning between individuals diagnosed with both social phobia and avoidant personality disorder and those diagnosed only with social phobia (p. 481). Thomas A. Widiger (1992) analyzed three studies on social phobia and avoidant personality disorder. All three studies "did find significant differences between the [social phobia] subjects with and without [avoidant personality disorder]," but Widiger noted that the key takeaway from the studies was that there were relatively no "substantial or qualitative distinctions between [social phobia] and APD" (p. 341).

Given these conclusions, James D. Herbert, Debra A. Hope, and Alan S. Bellack (1992) published a paper discussing the validity of the distinction between social phobia and Avoidant personality disorder. They used questionnaires, such as the Social Phobia Anxiety Inventory, the Social Avoidance and Distress Scale, and the State-Trait Anxiety Inventory-Train Form, to measure subjects' psychopathology. They also assessed social skills using a role-play test and an impromptu speech. Based on the results of their study, Herbert, Hope and Bellack concluded that the distinction between social phobia and avoidant personality disorder fails in the three ways that disorders must differ: "conceptually distinct, discriminable from one another, and as mutually exclusive as possible" (p. 337); they conclude that the disorders "represent different points on a continuum of severity" (p. 338).

This is further supported by Maria Tillfors and Lisa Ekselius' findings in 2009, based on the *DSM-IV* descriptions of both disorders. They analyzed the overlap between social phobia and avoidant personality disorder, studies of treatment, familial and genetic studies, and longitudinal studies. Though they explicitly state that more research is needed in order to support a "continuum hypothesis," they do report that there is an "extremely high degree of overlap between social phobia and avoidant personality disorder" (p. 31). The most notable differences reported in this paper are the responses to cognitive behavioral therapy and levels of impairment, as was also shown in the other studies discussed above.

This spectrum-based model has been integrated into the *DSM-V* in the diagnosis of autism spectrum disorder. In the fourth edition of the *DSM*, patients could be diagnosed with autistic disorder, Asperger's disorder, childhood disintegrative disorder, or pervasive developmental disorder not otherwise specified (APA, 2013a). With an increase of research-based knowledge about the four pervasive developmental disorders, the *DSM-V* contributors created a single umbrella disorder that they believe "will improve the diagnosis of [autism spectrum disorder] without limiting the sensitivity of the criteria, or substantially changing the number of children being diagnosed" (APA, 2013a). This same logic could potentially be applied to avoidant personality disorder and social phobia since the diagnostic criteria are so similar, and individuals diagnosed with avoidant personality disorder are generally also diagnosed with social phobia.

These criticisms lend to the argument that avoidant personality disorder is a socially constructed disorder. Social constructionism is a view of the world that emphasizes communal interchange. Paul A. Boghossian of New York University's Department of Philosophy describes social constructed concepts as those that, "had we been a different kind of society, had we had different needs, values or interests, we might well have built a different kind of thing, or built this one differently" (p. 1). The way that we view, describe, and even experience the world is dependent on our social selves. As Kenneth J. Gergen (1985) describes, this view calls into question pre-existing notions about mental disorders: "professional agreements become suspect; normalized beliefs become targets of demystification; 'the truth' about mental life is rendered curious....the contemporary views of [psychology] on matters of cognition, motivation, perception, information processing, and the like become candidates for historical and cross-cultural comparison" (p. 271). In a field as scrutinized as that of mental health, this can be especially concerning as a potential de-legitimizing critique.

The notion of social constructivism is even more pertinent when analyzing personality disorders. As previously discussed, individuals diagnosed with personality disorders are disparagingly regarded by mental health professionals. While other mental disorders may be considered treatable, personality disorders are not, due to their ego-syntonic nature. This means that the symptoms that these individuals present are nearly inseparable from the individual and reflect that individual's self-image. As the name of the category suggests, the symptoms of personality disorders are pervasive enough to become synonymous with the individual presenting them.

When looking specifically at avoidant personality disorder, it is possible that this disorder is an exacerbated version of social phobia categorized separately to emphasize the more debilitating nature of the disorder. In Brown, Heimberg, and Juster's (1995) study comparing social phobia and avoidant personality disorder, they measured the severity of an individual's impairment based on the *DSM* criteria for the disorders using four questionnaires: the Social Avoidance and Distress Scale, the Fear of Negative Evaluation Scale, the Social Interaction Anxiety Scale, and the Social Phobia Scale. On all four of these questionnaires, participants with a dual diagnosis of social phobia and avoidant personality disorder scored higher (22.8, 26.9, 56.1, and 39.9 respectively) than participants diagnosed only with social phobia (20.6, 25.1, 50.9, and 32.1 respectively). These results, supported by the other research discussed previously, support the theory that these disorders are really one and the same, with avoidant personality disorder exemplifying the same symptoms in a more severe way.

Due to the more extreme presenting issues in individuals with a diagnosis of avoidant personality disorder, it is possible that this intense diagnosis was created in order to separate those whose symptoms are considered manageable from those whose are more difficult to manage. As previously mentioned, patients with personality disorders are judged more harshly by mental health professionals than patients with other diagnoses. In her article debating the affects of diagnoses, Noriko Ishibashi (2005) discusses the consequences of naming a person's symptoms and giving them a diagnosis. She asserts that "people attach additional meanings once they understand that an individual has been given a certain diagnosis. Their conception of the individual is confined by the individual's categorization, or diagnosis" (p. 69). Once a patient is diagnosed with a personality disorder, it immediately changes the way he or she is seen by the world. Because these disorders are seen as chronic, as opposed to fatal or curable, this perception is also chronic. They are often self-fulfilling prophecies, causing the individual to regress further into their symptomatic behavior as they live up to the "expectations" that accompany their diagnosis.

This is most certainly applicable to a diagnosis of avoidant personality disorder. Not only may mental health professionals who may be trying to treat avoidant symptoms regard them as untreatable, the patients themselves may believe that they are stuck. This is partially attributable to the ego-syntonic nature of personality disorders, but is also due to the pathology of the disorder. As a chronic disorder, symptoms are apparent throughout childhood and are diagnosed in adolescence or early adulthood. An individual diagnosed with avoidant personality disorder does not know him or herself without his or her symptoms, and may not know him or herself without his or her diagnosis.

Because a diagnosis of social phobia is regarded as more treatable than a diagnosis of avoidant personality disorder, a change in the *DSM* diagnosis could lead to significant improvements in treatment and self-awareness in patients currently diagnosed with avoidant personality disorder. One possible change is combining the diagnoses into a single umbrella disorder, like the *DSM-V* did with autism spectrum disorder, as previously mentioned. Another option is the deletion of the diagnosis of avoidant personality disorder and the addition of severity or course specifiers in the coding of social phobia. An example of this can be seen in the *DSM-V* diagnostic criteria for major depressive disorder. The diagnosis can be made more specific by indicating its severity (mild, moderate, severe, or with psychotic features) or course (in partial remission or in full remission; APA, 2013b, p. 162). By including this coding feature, individuals currently diagnosed with avoidant personality disorder can have a "less severe" diagnosis of social phobia, which is seen as more manageable by patients and mental health professionals alike.

Due to the extreme overlap between the *DSM-V* criteria for social phobia and avoidant personality disorder, it is likely that they are two manifestations of the same condition. It is possible that the diagnoses have been differentiated due to stigma on the part of mental health service providers as well as the general public. This socially constructed diagnosis does more harm than necessary, negating its intended purpose of being descriptive and facilitating more reliable diagnoses (APA, 2013b, p. xli). By maintaining two separate diagnoses, the *DSM-V* is contributing to the stigmatization of individuals diagnosed with personality disorders and may even be inhibiting a person's success in treatment.

References

American Psychiatric Association. (2013b). *Autism Spectrum Disorder* [Fact sheet].

American Psychiatric Association. (2013b). *Diagnostic and statistical manual of mental disorders* (5th ed.). Washington, DC.

Angell, M. (2011, June 23). The Epidemic of Mental Illness: Why?. *The New York Review of Books*.

Boghossain, P. A. (n.d.). What is Social Construction? Retrieved from http://philosophy.fas.nyu.edu/docs/IO/1153/socialconstruction.pdf

Brown, E. J., Heimberg, R. G., & Juster, H. R. (1995). Social Phobia Subtype and Avoidant Personality Disorder: Effect on severity of social phobia, impairment, and outcome of cognitive behavioral treatment. *Behavior Therapy, 26*, 467-486.

Frances, A. (2010, June 2). Psychiatric Fads and Overdiagnosis: Normality is an endangered species. *Psychology Today*.

Gergen, K. J. (1985). The Social Constructionist Movement in Modern Psychology. *American Psychologist, 40*(3), 266-275.

Gray, K. (2013, March 18). Are we over-diagnosing mental illness?. *CNN*.

Herbert, J. D., Hope, D. A. & Bellack, A. S. (1992) Validity of the Distinction Between Generalized Social Phobia and Avoidant Personality Disorder. *Journal of Abnormal Psychology, 101*(2), 332-339.

Hill, T. (2014). Understanding the Avoidant Personality. *Psych Central.*

Ishibashi, N. (2005). Barrier or Bridge? *Smith College Studies in Social Work, 75*(1), 65-80.

Lewis, G. & Appleby, L. (1988) Personality disorder: the patients psychiatrists dislike. *The British Journal of Psychiatry, 153.* 44-49.

Long, P. W. (2015). Avoidant (Anxious) Personality Disorder. *Internet Mental Health.*

Mientka, M. (2013, May 12). Is Mental Illness Over-diagnosed? Backlash over the new DSM-V. *Medical Daily.*

National Institute of Mental Health (2007). *Avoidant Personality Disorder* [Fact sheet].

National Institute of Mental Health (2007). *Personality Disorders* [Fact sheet].

Skodol, A. E., Gunderson, J. G., Shea, M. T., McGlashan, T. H., Morey, L. C., Sanislow, C. A., Bender, D. S., ...Stout, R. L. (2005, October). The Collaborative Longitudinal Personality Disorders Study (CLPS): Overview and implications. *Journal of Personality Disorders, 19*(5), 487-504.

Tillfors, M. & Ekselius, L. (2009). Social Phobia and Avoidant Personality Disorder: Are they separate diagnostic entities or do they reflect a spectrum of social anxiety? *Israel Journal of Psychiatry & Related Sciences, 46*(1), 25-33.

Turner, S M., Beidel, D. C., Dancu, C. V., & Keys, D. J. (1986). Psychopathology of Social Phobia and Comparison to Avoidant Personality Disorder. *Journal of Abnormal Psychology, 95*(4), 389-394.

Turner, S. M., Biedel, D. C., & Townsley, R. M. (1992). Social Phobia: A comparison of specific and generalized subtypes and avoidant personality disorder. *Journal of Abnormal Psychology, 101*(2), 326-331.

U.S. Department of Health and Human Services. (2013). *Results from the 2012 National Survey on Drug Use and Health: Mental health findings* [Data set].

Widiger, T. A. (1992). Generalized Social Phobia Versus Avoidant Personality Disorder: A commentary on three studies. *Journal of Abnormal Psychology, 101*(2), 340-343.

A Social Disease: Examining Autism Spectrum Disorder through a Social Constructionist Lens

Laura Nessler

Even before its addition to the *Diagnostic and Statistical Manual of Mental Disorders* (DSM) in 1980, autism (formerly known as autistic disorder and now known as autism spectrum disorder) has been a diagnosis fraught with controversy. While autism's origins and apparent increase in prevalence are two of its most well-known controversies, another question emerges from a social constructionist standpoint: In what sense does autism exist at all? My question is not *if* autism exists, but instead how it has emerged as a diagnosis, how it functions, and whose interests this diagnosis serves.

I will argue that the diagnosis of autism spectrum disorder is socially constructed by the neurotypical in order to explain behavior that differs from their norms. Within the neurotypical community, various experts and neurotypical parents of autistic children have made especially significant contributions to this diagnosis. Neurotypical experts in the fields of psychiatry, speech therapy, education, and others construct autism as a deficit that must be fixed, while many neurotypical parents construct this deficit as separate from their "real" child. However, with the growth of the neurodiversity movement, autistic individuals are starting to construct their own definitions of autism based on their own experiences rather than on how they affect the neurotypical.

In order to understand this construction of autism, it is necessary to define what exactly is meant by "neurotypical" and "neurodiverse." The term "neurodiverse" emerged about ten years ago as an alternative way of differentiating those with autism from the "neurotypical," or those without autism. According to Thomas Armstrong, the neurodiversity movement emerged as an autism rights movement, or a desire to be seen as "different rather than disabled" (2010, p.7). The term neurodiverse has since been embraced by other groups, including dyslexics and those with Down's Syndrome (Armstrong, 2010); however, in this paper I will use the word "neurodiverse" to identify those with autism spectrum disorder exclusively. Similarly, even though Armstrong argues that there is no such thing as a truly "neurotypical" brain since every brain is unique, I will use the word "neurotypical" to refer to those who do not meet the criteria for autism spectrum disorder. Although human behavior exists along a continuum, the real and perceived differences between the "neurotypical" and the "neurodiverse" are important because they are the basis upon which the autism spectrum disorder diagnosis has been constructed.

Kenneth J Gergen defines social constructionist inquiry as "concerned with explicating the processes by which people come to describe, explain, or otherwise account for the world (including themselves) in which they live" (1985, p. 266). Social constructionists believe that knowledge cannot be objective, but rather is constructed through forms of social exchange that are influenced by their circumstances. As Gergen states, social constructionism has been used to challenge the "objective basis" of many psychological concepts from schizophrenia to suicide. I believe the diagnosis autism spectrum disorder is especially "circumscribed by culture, history, and social context" as well (1985, p. 267).

According to the fifth edition of the *Diagnostic and Statistical Manual of Mental Disorders* (DSM-5), the essential features of autism spectrum disorder are "persistent impairment in reciprocal social communication and social interaction, and restricted, repetitive patterns of behavior, interests, or activities" (American Psychiatric Association, 2013, p. 53). However, the definition of autism spectrum disorder has changed since the word "autism" was coined in 1940. This change illustrates how creating a diagnosis is an inherently social process affected by changes in society. One recent and dramatic example of this change was the shift from the three separate diagnoses of autistic disorder, Asperger's disorder, and pervasive developmental disorder not otherwise specified to a single diagnosis of autism spectrum disorder. This shift occurred between the publication of the DSM-IV Text Revision (DSM-IV-TR) and the DSM-5. Most noticeably, this change essentially eliminated Asperger's disorder from the DSM-5. Other changes occurred as well, such as the stipulation that to be diagnosed with autism spectrum disorder, one must meet three out of three deficits in communication and social interaction. Previously, autistic disorder and Asperger's disorder were defined by a more complicated combination of various factors.

Since the redefinition of autism in the DSM-5 was the result of social exchange and negotiation, it is impossible to pinpoint a singular force in society that elicited this change. One way of explaining this negotiation described by Hacking (as cited in Valentine, 2010) is the "looping effect." This looping effect summarizes the relationship between experts and parents in the social construction of autism. The loop begins when parental activism raises awareness of a disease and increases the availability of services for those with the disease. Since such services are contingent upon one being diagnosed with the disease, difficult situations arise when a child is "just outside" the criteria for the disease, which leads to a broadening of the criteria so that fewer children are left without services. As more children are diagnosed, awareness increases even more, and the loop continues. This loop underscores how autism is not an objective reality but rather a label that is malleable to social forces and opinions.

Although changes to the diagnosis of autism are generally created by the neurotypical and chiefly affect the neurodiverse, Linton, Krcek, Sensui and Spillers (2014) conducted an online survey of opinions from the neurodiverse about the changes to the autism diagnosis proposed in the DSM-5. Some self-identified autistics believe the new DSM-5 criteria is an oversimplification of autism, while some self-identified individuals with Asperger's are more concerned with the elimination of the term. There were also respondents who believed that the new criteria for autism will help improve accuracy of diagnosis. However, a common theme of distrust regarding neurotypical definitions of their identities ran throughout these discussions. Those who were in favor of the DSM-5 changes recalled experiences where they were misdiagnosed or misunderstood by neurotypical doctors under the old DSM-IV-TR guidelines. One respondent's comment sums up this distrust of neurotypical experts: "That's why I say that 'experts' frequently don't get autism at all. They truly need our input. However, I also realize that they do not generally consider us a reliable source" (Linton, et. al, 2014, p. 72).

Another respondent pointed out how the DSM-5's "oversimplification" of Asperger's disorder (now subsumed into autism spectrum disorder) specifically serves the interest of neurotypical medical professionals: "AS [Asperger's Syndrome] is not simple...There are things about it that are extremely difficult to understand, and creating an apparently simpler means of dxing [diagnosing] does nothing but make doctors jobs easier" (Linton, et. al, 2014, p. 73). This observation parallels Gergen's social constructionist principle that a concept's ability to last over time does not have to do with its empirical validity as much as with the "vicissitudes of social process" (1985, p. 268). Thus, an idea's merit is intrinsically linked to its comprehensibility or ability to be communicated to others. It follows that a simpler definition of autism serves the interests of the diagnosis itself since it enhances its ability to be communicated and thus its longevity.

Despite their disagreements on the validity of the new autism spectrum disorder diagnosis, many neurodiverse individuals nevertheless recognize the importance of neurotypicals' constructions of their identities. One survey respondent who identified as having Asperger's was concerned about being "lumped together" with those who a have "full on" autism. This respondent said: "...I think my biggest fears are that Aspies [those with Asperger's] will be treated a bit like their stupid. Not saying that the fully autistic are, just saying that that is how the general NT's [neurotypicals] see them/us" (Linton, et al., 2014, p. 74). This respondent's concern underscores the power of the neurotypical to construct the identity of the neurodiverse. As Linton et al., note, this particular concern is based on stereotype rather than fact. However, stereotypes can still exert tremendous power over a group. Even though stereotypes thrive on ignorance, they can affect anyone, including those who provide the neurodiverse with medical care and other services.

While psychiatrists were largely responsible for the changes within the DSM-5, they are not the only experts who play a role in constructing the diagnosis of autism spectrum disorder. Within the realm of education, Molloy and Vasil (2002) argue that Asperger's disorder was socially constructed because of its value to special education. Even though Asperger's disorder has now been incorporated into autism spectrum disorder, Molloy and Vasil's thesis is still applicable to this new diagnosis. According to Molloy and Vasil, special education is a way of changing a child's unconventional behaviors so that they better fit into the mainstream education system. This convention ensures that deficits are interpreted as within the individual rather than within the educational system at large. Molloy and Vasil also point out that Asperger's disorder provides a common label for experts from different disciplines, such as teachers, speech therapists and occupational therapists. While a label like Asperger's disorder or autism spectrum disorder creates a common meaning across disciplines, it also implies a specific action, or treatment, within each discipline. As Gregen states, "descriptions and explanations form integral parts of various patterns. They thus serve to sustain and support certain patterns to the exclusion of others" (1985, p. 269). This statement's applicability to autism can be understood by replacing the word "patterns" with the word "treatments." For example, every time a speech pathologist constructs a client as autistic they are reinforcing an established way of treating that client based on the norms of their discipline.

Of course, since the meaning of autism is constantly being reconstructed, treatments for autism (even within the same discipline) vary widely as well. The language in the DSM-5 is vague enough to allow neurotypical experts to construct autism in a way that aligns with their treatment philosophy. However, despite its vagueness, the DSM is extremely clear about framing autism as a condition that requires treatment, even though it leaves up to the experts what that treatment should be. For example, all three of the diagnostic criteria in Criterion A start with the word "deficit." These deficits must occur in the areas of social-emotional reciprocity, nonverbal communicative behaviors and in developing, maintaining and understanding relationships (American Psychiatric Association, 2013, p. 50). The word "deficit" itself implies that it is necessary to add something where too few of something exists. The word implies that neurodiverse individuals are "incomplete" versions of neurotypical people because of their lack of social skills.

Since neurotypical "experts" directly benefit financially from treating those with autism spectrum disorder, it is not difficult to imagine their role in socially constructing this disorder according to their interests. However, because of the taxing nature of autism spectrum disorder upon parents, it is difficult to imagine how the construction of an autism spectrum disorder diagnosis might serve parents' interests. As Valentine (2010) points out, autism often has "a nightmarish quality...children [with autism] refuse to be touched, fail to communicate, engage in apparently bizarre behaviour or scream for hours at the slightest change to routine, or at loud noises or bright lights" (p. 956). However, it is important to differentiate an autistic child's behavior from their diagnosis of autism itself. While dealing with a child who acts in way foreign to many neurotypical parents is a disconcerting experience, parents can construct the diagnosis itself in a variety of different ways. Although each parent's construction of the diagnosis is different, most neurotypical parents' constructions of the diagnosis revolve around how autism is a disease separate from their child that can therefore be treated, and perhaps someday even cured.

Avdi, Griffin, and Brough (2000) conducted a discourse analysis on interviews of parents of autistic children that illustrates this point. The interviews addressed how parents construct "the problem" of their child being diagnosed with autism spectrum disorder. The analysis conducted on these interviews shows that parents construct the diagnosis using three different discourses. The first discourse that Avdi, et. al found was the "discourse of normal development." In this discourse, parents described how they compared their autistic children to the idea of a "normally developing" child. One interviewee used this discourse by saying, "normally your child is born and it goes through the various stages and it sort of just grows up and [laughs] does what it should do at the right times" (Avdi, et. al, 2000, p. 245-246). This discourse underscores how parents compare their autistic children not just to other children, but to expectations they had for their own children. This normalcy is a major component of the child that many neurotypical parents think was "stolen" from them by autism.

The second discourse that Avdi et al. observed in their study was the medical discourse. This discourse posits that an autism diagnosis, for all of its difficulties, also comes with a sense of relief because as a diagnosis, there is a known course and could someday be a cure. Many parents who used the medical discourse derived comfort from having a label for their children's difficulties that might otherwise be difficult to explain. The medical discourse also most explicitly gives parents a way to separate the symptoms of a disease from who they perceive their child to really be. Avdi et al. noted that for many parents, "Autism was constructed as an 'enemy from within,' something constantly to watch out for, scrutinize, analyse and challenge" (2000, p. 248). Thus, when parents were battling this "enemy within," they saw themselves as battling with autism, not with their children themselves.

The third discourse of this study was the discourse of disability, which focused on how children would be stigmatized because of their diagnosis. Although these parents feared that their children would be stigmatized, some parents stigmatized others in an effort to construct their child as more "normal." For example, one interviewee said that he would rather his autistic child go to a "mainstream nursery" because he believed the following:

"He is better off mixing with, it sounds terrible really to say, but what I consider to be   more normal children than him and if he's going to learn from them and pick up their ways and things like that I'd rather him be in that environment than in one where there are people like him, if you like, and worse in some cases" (p. 251).

This passage illustrates that this parent believes that autism is a condition that (to some extent) can be influenced by environment. With enough "normal" influence from others, perhaps the "normal" child within his son will emerge, thereby beating out the autism. If this parent believed that autism was intrinsic to his son's identity, he would not be concerned about his environment relative to how other "people like him" could influence him.

By constructing autism as a diagnosable disease, a great deal of pressure is put upon parents to pick the right treatment that will rescue their child who has been "taken" by autism. Kylie Valentine (2010) conducted a qualitative study of parents' experiences of an autism diagnosis and treatment that speaks to this pressure. Valentine was chiefly concerned with what role choice plays in treatment options for parents.   She found that treatment choice was neither universally beneficial nor universally detrimental to parents. Some parents felt unqualified yet obligated to choose a course of treatment for their autistic child, while others became fully engaged in their child's treatment and advocacy for others (Valentine, 2010). However, regardless of their feelings on choice, the parents interviewed all felt intense pressure to reclaim their children from autism. One parent who was overwhelmed and disempowered by all of the treatment choices said, "...We're just terrified we're going to make the wrong decision" (Valentine, 2010, p. 955). Even though autism is not a life-threatening illness in the literal sense, parents' fear of failure surrounding treatment makes it seem as if their children's lives are at stake. Valentine notes that both parents empowered and disempowered by choice share the sentiment that "it's up to the parents" (2010, p. 956).

Up until this point, I have discussed the asymmetrical relationship between the neurotypical and the neurodiverse with respect to how the diagnosis of autism has been socially constructed. The neurodiverse receive the label of autism spectrum disorder while the construction of the term is largely based on the opinions and interests of the neurotypical. However, as Gergen states, "knowledge is not something people possess somewhere in their heads, but rather, something people do together" (1985, p. 270). Therefore, even though the diagnosis of autism was constructed together by one group of people, this diagnosis is not a static, constant entity once it has been constructed. Instead, the diagnosis is malleable to other groups who are able to construct knowledge together, including the neurodiverse themselves.

Many different constructions of autism spectrum disorder are subsumed under the umbrella of the neurodiversity movement. It could even be argued that just as every brain is different, so is every person's construction of autism. One self-reported autistic person said about the validity of the autism diagnosis, "it is not valid to have a diagnosis that caters to prejudice" (Linton, et al., 2014, p. 73). Thus, there are some neurodiverse individuals who do not believe that autism should be a diagnosis at all.

Thomas Armstrong quotes an autistic woman, Amanda, who highlights the double standard inherent in treating autism as a disorder based on deficits. Amanda communicates with objects with the same frequency and fluency that the neurotypical communicate with people. In an online video called "In my Language," she demonstrates what this communication looks like and says through a voice synthesizer on the computer, "I find it very interesting...the failure to learn your language is seen as a deficit, but failure to learn my language is seen as so natural" (2010, p. 54). This observation underscores the social constructionist principle that language and social interchange are how people categorize and make sense out of the world. Autistics have varying degrees of verbal language and social skills. These impairments do not only make up the diagnosis of autism spectrum disorder, they also put autistic people at a disadvantage when it comes to having a say about how autism spectrum disorder is constructed.

However, as Armstrong (2010) notes in the case of Amanda and others, the internet has given many neurodiverse individuals a way to unite and communicate. The internet has also given the neurodiverse the ability to co-construct their identities in a way they hadn't been able to before. The growth of the movement has also lead to a more nuanced and diverse view of neurodiversity itself. A 2013 study conducted by Kapp, Gillespie-Lynch, Sherman, and Hutman about the awareness of the neurodiversity movement in relationship to opinions about autism reflects this multifaceted view of neurodiversity.

Some of the findings from Kapp et al.'s study were unsurprising. For example, autistic people were, as a whole, less interested in finding a cure for autism or determining its causes than parents of autistic children were. Yet, the study ultimately came to the conclusion that "awareness of neurodiversity and self-identification as autistic correspond with a deficit-as-difference conception of autism" (Kapp, et al., 2013, p. 66). This "deficit-as-difference" model supports an emphasis on the strengths of the neurodiverse, and is not interested in curing the disease. However, it also acknowledges that certain services can help the neurodiverse adapt to a neurotypical-dominated world.

This shift to the "deficit-as-difference" model in the neurodiversity movement is similar to a recent shift in the social model of disability. The neurodiversity movement itself is heavily influenced by the social model of disability, which began gaining traction shortly before the advent of the neurodiversity movement. The social model of disability, which is influenced by social constructionism, believes that impairment itself exists objectively in the world, whereas "disability," or how we treat these impairments, is socially constructed (Molloy & Vasil, 2002). Molloy and Vasil note that there has been a second wave of writing in disability studies that has questioned the split between "impairment" and "disability" in the social model. There is concern that this split ignores and undermines the lived experiences of those with impairments. The focus becomes too much on the social construction of disability rather than on the people with the impairments. Those who are concerned with this split argue that disability studies should "refocus attention onto impairment in an attempt to redress the abandonment of the body that is the legacy of the social model" (Molloy & Vasil, 2002, p. 663).

This distinction between the first and second waves of the social disability model is reflective of the evolution taking place in the neurodiversity movement as well. By focusing too much on the social construction of the autism diagnosis, one loses a sense of the needs and experiences of the neurodiverse themselves. As Kapp et al. found in their survey, those with autism who support the neurodiversity movement are not necessarily opposed to services that will help them fit into a neurotypical world (2013). While proud of their differences, these neurodiverse individuals recognize that their lives will be easier if they learn certain skills valued by the neurotypical. Thus, the experience of the autistic person whose life is improved by these services and treatments should not be constructed as counter to the neurodiversity movement. Just as the social model of disability is adjusting its thinking to encourage those with disabilities to talk about their impairments, the neurodiversity movement is starting to accept that autism can be an impairment at the same time that it is a difference or strength.

By looking at autism spectrum disorder through a social constructionist lens, it becomes apparent that there are many different ways of "seeing" the diagnosis that differ depending on one's vantage point. Mental health and medical experts construct the disorder in a distinct way, as do parents of autistic children, and those with the diagnosis itself. Until recently, the former two groups have had a greater influence on the construction of the diagnosis, but the rise of the neurodiversity movement signals a change in how and by whom autism spectrum disorder is constructed. As Gergen points out about social constructionism, this is not to say that "anything goes" when it comes to defining autism, or that the diagnosis itself holds no meaning because of the different things it means to different people. Instead, I believe that these different perspectives provide a richer understanding of the condition overall. Perhaps engaging with this richer understanding of autism spectrum disorder will help the neurodiverse and the neurotypical live together in a more mutually beneficial way in the future.

References

American Psychiatric Association. (2013). Diagnostic and statistical manual of mental disorders (5th ed.). Washington, DC: Author.

Armstrong, T. (2010). *Neurodiversity: Unleashing the advantages of your differently wired brain.* Da Capo Press: Cambridge, MA.

Avdi, E., Griffin, C., & Brough, S. (2000) Parents' construction of the 'problem' during assessment and diagnosis of their child for an autistic spectrum disorder. *Journal of Health Psychology, 5*(2), 241-254.

Gergen, K.J. (1985). The social constructionist movement in modern psychology. *American Psychologist, 40*(3), 266-275.

Kapp, S. K., Gillespie-Lynch, K., Sherman, L.E., & Hutman, T. (2013). Deficit, different or both? Autism and neurodiversity. *Developmental Psychology, 49*(1), 59-71.

Linton, K.F., Krcek, T.E., Sensui, L.M., & Spillers, J.L.H. (2014) Opinions of people who self-identify with autism and Asperger's on *DSM-5* criteria. *Research of Social Work Practice, 24*(1), 67-77.

Molloy, H., & Vasil, L. (2002). The social construction of Asperger syndrome: the pathologising of difference? *Disability & Society, 17*(6), 659-669.

Valentine, K. (2010). A consideration of medicalisation: Choice, engagement, and other responsibilities of parents with autism spectrum disorder. *Social Science & Medicine, 71*, 950-957.

A Feminist Analysis of Agoraphobia

Marguerite Barrett

Agoraphobia is an anxiety disorder that impacts the lives of approximately five percent of Americans each year. Fundamentally, the word can be broken into two distinct Latin root words "agora" meaning place of assembly and "phobia" meaning fear. However, agoraphobia runs deeper then one simply being afraid of public places, and there many complex components that accompany this anxiety disorder. Statistics show us that while agoraphobia affects five percent of all people in the United States each year research suggests nearly two-thirds of those people are women (American Psychiatric Association, 2013). Other studies propose even higher numbers suggesting women are three times as likely to be diagnosed, and a rate of recurrence double to that of men (Yonkers, Zlotnick, Allsworth, Warshaw, Shea, & Keller, 1998). In this analysis feminist theory will be used as a lens with which to understand this disorder's vast gender disparity. While the primary goal is not to minimize the impairment males with this disease can face, the focus of this critique will be on why women are more likely to be diagnosed with this disorder by looking beyond a biological approach and viewing agoraphobia within a wider social context.

Diagnostic Criteria

The most recent publication and fifth edition of the *Diagnostic and Statistical Manual of Mental Disorders* (DSM-V) outlines several key criteria necessary in order to make a diagnosis of agoraphobia. A person with agoraphobia must have a marked fear of two or more of the following situations; using public transit, being in open spaces, being in enclosed spaces, standing in line or crowd, or being outside of the home alone (American Psychiatric Association, 2013). A person with agoraphobia could experience extreme fear and anxiety about anything from riding the bus, driving across a bridge, to going to the supermarket, and each person with agoraphobia will have his or her own unique specific set of fears to accompany his or her experiences. Another commonality and additional DMS-V criterion for diagnosis is that the individual fears or avoids the above listed situations due to the belief that escape might not be possible, or that help may not be readily accessible in the event of experiencing panic like symptoms, or otherwise embarrassing or incapacitating symptoms. These situations practically always incite fear and anxiety and thus are actively avoided. A person with agoraphobia might avoid the agoraphobic situation altogether, require the need for a companion, or undergo the situation but with profound fear. Active avoidance, which can be defined as behaving in ways that are purposefully aimed to avert or lessen exposure to agoraphobic situations, is an essential feature in diagnosis (American Psychiatric Association, 2013).

Furthermore, the agoraphobic situations are considered to be abnormal fears within a sociocultural context (American Psychiatric Association, 2013). For instance a person with agoraphobia might fear going to the grocery store, even if it is a nice day in a safe area, something typically considered to be a rather basic mundane task to the average person, and not a situation worthy of fear or panic. Additionally, in the event that one does have another medical disease present such as Crohn's disease, the fear and avoidance are still clearly extreme. One major diagnostic indicator of this disorder is that there is marked impairment in an area of significant functioning. Some examples include; if you cannot go to work because you are afraid to take the bus there, if you do not seek medical attention for an illness because you are fearful to leave the house alone, if you fail to attend an important funeral or social event because the thought of taking a plane is too distressing, or if you go without food because the grocery store seems too terrifying, etc. Finally, these symptoms need to be persistent, typically last for a period of over six months and cannot better be explained by another diagnosis (American Psychiatric Association, 2013).

Course and Comorbidity

According to the DSM-V approximately two-thirds of people with agoraphobia experience the initial onset before the age of thirty-five, with the highest risk of development during late adolescence and early adulthood. The course of agoraphobia is typically chronic, and unfortunately complete remission is rare, around ten percent, unless treated. As the severity of this disorder increases, the chance for recovery decreases with frequency of relapse increasing (American Psychiatric Association, 2013). Having agoraphobia has been noted to greatly increase one's chance to develop other mental disorders such as depression, dysthymia, or substance abuse disorder (American Psychiatric Association, 2013). In many cases the presence of another anxiety disorder may precede the onset of agoraphobia, while other comorbid conditions such as depressive and substance abuse disorders typically develop during the course of the disorder. This pattern could be attributed in part to the social isolation and functional impairment that accompany an agoraphobia diagnosis.

Living with Agoraphobia

People living with agoraphobia suffer from what could be considered a "fear of fear" (Worell, 2001). A person experiences a panic attack, which can lead to an irrational fear of experiencing those same panic-like symptoms again, and thusly causing people to avoid situations that could potentially incite those feelings. Agoraphobic individuals are afraid of becoming anxious (Carlson, K. J., Eisenstat, S. A., & Ziporyn, T. D., 2004). A primary facet of this disorder is the fear of these symptoms emerging in an uncontrollable way and manifesting in physical symptoms such as fainting, vomiting, or incontinence. The avoidance allows individuals to harness a sense of control over these feelings.

Socially, living with agoraphobia presents a wide array of challenges. Individuals with this disorder frequently struggle to develop a sense of self-sufficiency. Oftentimes they have to rely on a significant other to be with them most of the time (Worell, 2001) and frequently exhibit anxious attachment, or anxiety of not being taken care of. It is common for those living with agoraphobia to mislabel their emotions and label arousing emotions as panic. Furthermore, people with this disorder often struggle with maintaining employment, social isolation, marital troubles, and low self-esteem. Self-esteem is often directly correlated with how in control they feel: The less control they feel like they have over their lives and situations the lower their self-esteem is likely to be (Worell, 2011).

The DSM-V breaks down risks and diagnostic factors associated with agoraphobia into three categories: temperamental, genetic and physiological, and environmental. Each of these factors presents in many different ways and provides a unique challenge when understanding one's risk for developing agoraphobia. Biological diffidence, as well as neurotic temperament (examples include neuroticism and those prone to anxiety) both closely correspond with agoraphobia. However, these factors are also prevalent in many other anxiety disorders, such as panic, generalized anxiety, or phobic disorders. Furthermore, a characteristic of individuals dealing with agoraphobia includes the mentality that symptoms of anxiety can be harmful to them (American Psychiatric Association, 2013).

Childhood trauma, such as abuse or the death of a parent, are associated with the inception of agoraphobia. These instances may include, but are not limited to: separation, attack or mugging, sexual assault, or other such stressors. Individuals who have agoraphobia have distinguished that their "family climate" and "child-rearing behavior" were characterized by overprotection by guardian figures, and a decrease in warmth and nurturing (American Psychiatric Association, 2013).

According to the DSM-V, agoraphobia has the strongest association with the genetic factors that mark a proneness to phobias and anxieties. It also marks heritability for agoraphobia at 61% (American Psychiatric Association, 2013). This statistic in itself warrants potential critique and raises the question that social scientist so often face, the "nature versus nurture" debate. One could argue that simply boiling this complex illness down to a genetic or "nature" component is a rather reductionist approach. One must also consider how being raised or growing up around someone with this illness especially during the time of development and attachment could influence one's likelihood to develop the disorder themselves. Recent studies suggest that when it comes to mental illness it is not an either or debate at all, but rather that the development of such things requires a combination of a multitude of factors, or the idea of "nature via nurture" (Ridley, 2003). Genetic predispositions, biological composition, childhood development, life events, and social positions could all come into play in this case. In this analysis the focus will be on observing how one of these factors, social position, could have a major impact on the development of agoraphobia in women.

Feminist Theory

Agoraphobia has often been viewed as an issue among women. In 1949, a woman suffering with agoraphobia was described in the following way,

A woman who is chronically anxious and feared death or sudden illness if she went out on the street, on trains in cars, or to theatre, and the church. She had reached the point where she could not perform any of her duties and was helpless. Her husband has to remain home with her and even then she continues to be frightened. (Worell, 2001, p.110)

Now, more than half a century later, we see similar descriptions ascribed to women living with this disorder. Since that time women have seen more social, economic, and political rights in the Western world. However, feminist theory contends that many similar social pressures and standards for women are still ingrained within our socializing processes even today and that women remain a marginalized group within society (Sands & Nuccio, 1992).

There are multiple approaches that currently exist within feminist theory, and different approaches emphasize different components and facets of how women are oppressed. Feminist works throughout disciplines have furthered our understanding of the status of women within a patriarchal society, gendered bias in social and behavioral theories, and the feminization of poverty (Sands & Nuccio, 1992). Socialist feminism ties women's oppression to the intersection of sexism, racism, and class divisions, which are constructed through patriarchal capitalism (Coady, & Lehmann, 2007). In this sect of feminism there is an emphasis on organizing women to fight against gender specific aspects of oppression, such as issues of sexual abuse and reproductive rights. Liberal feminism highlights the disparity of political rights, opportunities, and justice within the existing political system (Coady, & Lehmann, 2007). The focus of liberal feminism is more centered on advocating for equal education, employment opportunities, and equal pay. Radical feminism finds patriarchy a pervasive influence that needs to be dismantled. They view the patriarchal power system as something that extends beyond just political spheres and is an integral part of our language, mass media, and everyday social interactions (Coady, & Lehmann, 2007).

While feminist theory is broad, and has many different sects and modalities of understanding, a commonality is that all seek to understand how social structures directly affect the lived experiences of marginalized and oppressed groups and assert that through being a woman within the context of an oppressive power structure there is a direct affect on one's life and experiences

(Swigonski, 1993). That is why we must look beyond medical science when it comes to disorders that affect women more prevalently and consider how they can be perpetuated through existing social contexts. This analysis will come from a standpoint that draws on different areas of a feminist perspective in order to understand the disparity in gender when it comes to this illness.

Feminist Critique

Recent research has stressed the role of trauma, rape, and possibly child sexual abuse in the histories of women who develop this disorder. These stressors are considered potential triggers for women to develop agoraphobia (Worell, 2001). Due to the high occurrence of these instances in women, with 17.7 million American women having been victims of attempted or completed rape and 7% of girls in grades 5-8 and 12% of girls in grades 9-12 saying they have been sexually abused (RAINN, 2015). It is common for women to be taught to actively avoid potentially dangerous situations: to not go to bars or nightclubs alone, to not walk around late at night by themselves, or jog with music playing, and to be cautious of the areas they choose to travel. Essentially, women are often already socialized in active avoidance and to be on "high alert" as a mechanism for protection and survival. One could argue that for agoraphobic women, this message has been internalized, distorted, and taken to irrational levels. That perhaps this fear women are taught they should have in certain situations cannot be turned off easily even when encountering situations that are considered safe within a sociocultural context. While studies show us that in reality, two-thirds of all assaults on women are committed by partners or someone they know, there is a broader fear and perception otherwise (RAINN, 2015). While agoraphobic women, like other non-agoraphobic women, may harbor this fear, it cannot be concluded that simply that fear alone is cause enough for agoraphobia, but rather this fear combined with other vulnerabilities and risk factors could increase the likelihood of developing this disorder.

As mentioned previously, a primary facet of the agoraphobic condition is the intense fear of "losing control". There is no set clinical definition for this term, and the experience is subjective to the unique experiences and fears of each individual, but it is a commonality between all of those living with agoraphobia. Agoraphobic individuals fear they may lose control over their bodily functions, fall unconscious, that their heart might stop, or that the panic might spiral so far out of control they lose control of their mind completely with no chance of ever regaining it. While these fears may seem irrational to those outside of the experience, they are part of the reality that agoraphobic patients have constructed for themselves. Studies show that for women there was a significant fear of becoming "completely hysterical" (Schmidt & Koselka, 2000). Moreover, studies show that women were more likely to have catastrophic thinking in regards to their panic experience compared to men (Schmidt & Koselka, 2000). Viewing this in both a historical and social context, the fear of losing control would be more common for women to experience, as the consequences for doing so have historically and socially been greater and more disparaging. Gendered insults such as a "crazy", "bitch", and "slut" all indicate a lack of control over one's self or desires, and are often used to invalidate a woman's actions or emotions. By imposing norms and values that seek to limit women's sexual, political, economic and social autonomy feminist theory asserts that oppressive patriarchal structures actively seek to control women (Coady & Lehmann, 2007).

Conclusions

In observance of these considerations we can seek to hone a greater understanding of the prevalent gender difference present within this diagnosis. These implications call for both qualitative and quantitative research to be conducted in regards to the development of this condition in women. As social workers and mental health professionals we should perpetually seek to increase our knowledge base and understanding of the diagnoses we ascribe clients, and not accept simple explanations for complex disorders. Furthermore, regardless of whether you approach treatment through a feminist perspective or other modalities an effort should be made to understand larger social issues that directly impact our client base on the marco level, so we can seek to provide the most effective treatments within our micro level practice.

References

American Psychiatric Association. (2013). Diagnostic and statistical manual of mental disorders: DSM-5. Washington, D.C: American Psychiatric Association.

Carlson, K. J., Eisenstat, S. A., & Ziporyn, T. D. (2004). *The new Harvard guide to women's health*. Cambridge, Mass: Harvard University Press.

Coady, N., & Lehmann, P. (Eds.). (2007). Theoretical perspectives for direct social work practice. 2nd ed. New York, NY: Springer.

Ridley, M. (2003). Nature Via Nurture: Genes, Experience, and What Makes Us Human. New York, New York: Harper Collins Publishing.

Sands, R. G., & Nuccio, K. (November 01, 1992). Postmodern Feminist Theory and Social Work. *Social Work, 37,* 6, 489-94.

Schmidt, N. B., & Koselka, M. (January 01, 2000). Gender Differences in Patients with Panic Disorder: Evaluating Cognitive Mediation of Phobic Avoidance. *Cognitive Therapy and Research, 24,* 5, 533-550.

Statistics | RAINN | Rape, Abuse and Incest National Network. (2015).

Swigonski, M. E. (June 01, 1993). Feminist Standpoint Theory and the Questions of Social Work Research. *Affilia: Journal of Women & Social Work, 8,* 2.

Worell, J. (2001). *Encyclopedia of women and gender: Sex similarities and diff*

erences and the impact of society on gender. San Diego, Calif: Academic Press.

Yonkers, K. A., Zlotnick, C., Allsworth, J., Warshaw, M., Shea, T., & Keller, M. B. (January 01, 1998). Is the course of panic disorder the same in women and men?. *The American Journal of Psychiatry, 155,* 5, 596-602.

Trauma Informed Assessment
of Clinical Presentation of Bipolar Disorder

Molly Feldheim

Bipolar I Disorder is a mental condition characterized by cyclical episodes of mania and depression. Episodes of mania involve high impulsivity, reactivity and elevated mood which may be accompanied by subsequent episodes of depression (American Psychiatric Association, 2013). Individuals diagnosed with bipolar disorder vary greatly in their clinical presentation of symptoms; these symptoms range from those with psychotic features such as hallucinations and delusions to those who experience frequent shifts from high to low mood (American Psychiatric Association, 2013).

The *Diagnostic and Statistical Manual of Mental Disorders* or DSM-V (American Psychiatric Association, 2013) lists the symptom criteria needed for a diagnosis of bipolar disorder. The DSM-V also states that a family history of bipolar disorder is one of the strongest predictors for a bipolar diagnosis, suggesting a genetic etiology. What the DSM neglects to report is the significantly higher rates of childhood trauma in individuals with a bipolar disorder diagnosis.

Trauma is defined as either (1) "unexpected, overwhelming intense emotional blows that assault the person from the outside and quickly become incorporated into the mind." (Terr 1990 as cited in Bloom, 1999) or (2) "when both internal and external resources are inadequate to cope with external stress" (Van der Kolk, 1989 as cited Bloom, 1999). Findings that suggest a heavy link between bipolar disorder and adverse childhood events have major implications for how practitioners understand the diagnostic category (Etain et al. 2008, Larsson et al. 2013).. In fact, when one attempts to understand individuals with an emphasis on their unique life experiences, a so-called 'trauma lens', most DSM diagnoses stop making much clinical sense at all. Trauma theory explains the long-term effects of trauma to the human psyche. These explanations are parallel to the diagnostic criteria for a bipolar diagnosis. Furthermore, a mental diagnosis may disempower and even re-traumatize people who seek out mental health services.

The purpose of this paper is to argue the inadequacy of the diagnosis of Bipolar Disorder, specifically using insights from trauma theory as well as to suggest a more nuanced understanding of individuals relevant to clinical assessment by attempting to show that the clinical presentation of bipolar disorder is not the result of a genetic deficiency but rather the effects of a traumatized psyche.

Psychological Trauma And Trauma Theory

The understanding of trauma in mental health professions has continued to evolve in light of new and emerging literature. Trauma theory came about in the 1970s in response to Vietnam War veterans returning to the US. The theory also addressed other survivor groups including holocaust survivors, refugees and victims of violent crimes such as domestic abuse or rape (Center For Nonviolence And Social Justice, 2014). This radically shifted the understanding of traumatized people from being thought of as sick, corrupted or deficient to being understood as someone "who has been injured and in need of care" (Center For Nonviolence And Social Justice, 2014). Most psychologists and other clinicians now agree that trauma does not lie in the external event itself, but in the individual's inner interpretation of the event. Traumatic events affect the entire person--their thoughts, emotions, sense of self, behaviors, relationships and how they view the world around them. Herman (1992) writes in her book, *Trauma and Recovery*, "Traumatic events are extraordinary, not because they occur rarely, but rather because they overwhelm the ordinary human adaptations to life" (as cited in Center for Nonviolence, 2014)

Trauma and Evolutionary Psychology

What Herman is referring to as "ordinary human adaptations to life" is our evolutionary responses to external stress. To understand trauma and the effects it has on our human psyche, trauma theory relies on a basic understanding of our 'mammalian heritage' (Bloom, 1999) of the fight-or-flight response and the need for social connections. As a species humans have been able to survive for so long in large part due to these evolutionary processes of fight-or-flight and social evolution. We developed into a social species to benefit from mutual protection. When our sense of protection is broken we rely on the fight-or-flight response (Bloom, 1999). Evolutionary psychology also reminds us that we are born with a set of innate emotions that allow us a set of predictable brain responses to external stress; this means that when we experience overwhelming emotions--such as with traumatic events--both our physical bodies and our psychological selves are damaged (Bloom, 1999). This is because our brains are designed to work as an 'integrated whole'--similar to a computer made up of many separate parts. When we experience trauma, we experience a fragmentation of ourselves and our brain function is altered (Bloom, 1999).

This flight-or-fight response is precisely what is altered in people who have been traumatized. When we sense danger our amygdala is activated, and we create new neural connections in response to the danger. Essentially, our brains change ever so slightly every time we are in fear for our lives. When children especially experience repeated threats and little protection, it has long-term effects on their flight-or-fight responses; even the slightest threat becomes overwhelming (Bloom, 1999).

Long Term Effects of Trauma

We know that traumatic events have a long-term effect on the human psyche. Trauma theory lists and explain several of these effects and how they arise from the traumatic experience. Besides the changes to fight-or-flight responses, these effects include but are not limited to, learned helplessness, an inability to modulate arousal, trouble with decision making under stress, emotional dissociation, and traumatic reenactment (Bloom, 1999). These changes are especially severe when trauma is repeatedly experienced in childhood. This is because children are still developing an ordinary adaptation to stress. When this development is interrupted by overwhelming emotional experiences it is derailed and altered to meet the emotional needs of the traumatized self (Center For Nonviolence And Social Justice, 2014).

People who have experienced trauma lose their ability to modulate arousal--meaning they are unable to calm themselves down or control the intensity with which they react to outside stimuli (Bloom, 1999). They often develop an all-or-nothing reaction; an on or off switch, with no in between. Children who have been traumatized are unable to develop a sense of emotional restraint; the importance or severity of a situation becomes irrelevant to their emotional response (Bloom, 1999). As a result, they tend to stay irritable, anxious, jumpy and on edge. They are unable to adjust these sensations to match the social setting. This absolutist thinking, irritability and anxiety are all often associated with a bipolar diagnosis, (American Psychiatric Association, 2013) which will be discussed below.

There is a clear importance of flight-or-fight for emotional survival and the impacts trauma has on this evolutionary process. As a result the traumatized individual is less likely to think clearly and make informed decisions under stress--mild or severe (Bloom, 1999). When we are under stress our decisions become based on impulse rather than diligence--action rather than thought. This type of reactivity is beneficial when we are in danger. However the traumatized person who always perceives danger will always make these types of rash decisions. Their perceived need for self-protection becomes more powerful than the need for self-regulation or impulse control (Bloom, 1999). Other thought patterns learned through the traumatic experience include absolute thinking, anger, and denial of personal difficulties (Bloom, 1999). Again, this looks similar to the diagnostic criteria discussed in the next section of this paper. For now let's explore two more effects of trauma that are hallmarks of Trauma Theory: dissociation and trauma reenactment.

Dissociation: Dissociation is the most extreme form of coping to avoid emotional danger. Dissociation is defined as "a 'disruption in the usually integrated functions of consciousness, memory, identity, or perception of the environment." (Bloom, 1999). It involves a splitting in the brain. Forms of dissociation may be anything from repressing a memory to separating the memory and the feelings associated with it, in order to make the memory less painful (Bloom, 1999). Trauma theorists call this coping mechanism 'emotional numbing.' Sometimes this numbing is used to survive repeated and continuous trauma experiences. Other times it is a way to cope in the aftermath of a traumatic event. Over time the traumatized person becomes better and better at numbing emotions which lends itself to many issues in their personal and social lives. They may avoid any relationship or social situation that may trigger traumatic memories or feelings associated with that memory (Bloom. 1999).

Trauma reenactment: perhaps one of the most perplexing and interesting aspects of trauma recovery is what trauma theorists refer to as 'trauma reenactment.' It is contradictory but not untrue that although traumatized persons will avoid triggering situations, they simultaneously crave those feelings and thoughts associated with their trauma. They will subconsciously find ways to reenact their trauma in their everyday lives (Bloom, 1999). Most of the time this means splitting the memory and emotions off into non-verbal images and sensations:

> We reenact our past everywhere- at home, at school, at the work-place, on the playground, in the streets. We cue each other to play roles in our own personal dramas, secretly hoping that someone will give us a different script, a different outcome to the drama, depending on how damaging our experiences have been. The cure is in the disease (Bloom 1999)

People who have suffered trauma are constantly trying to tell their story through their interactions with others. This storytelling can be manifested overtly or more disguised. Most of the time this expressivity comes from the subconscious (Bloom, 1999).

Bipolar: Trauma Lens Vs. Diagnosis

Despite knowing what we know about the long-term effects of trauma on individuals, it seems there is a profound lack of interest in trauma when it comes to clinical assessment of consumers. A 'trauma lens,' so named in this paper because of its impact on how practitioners see everything in relation to the individual, could be better understood as trauma-informed assessments. Most people receiving mental health services will receive a DSM diagnosis. The label of Bipolar disorder especially has become increasingly popular among mental health professionals (Geller & Luby, 1997). Consumers would greatly benefit from reframing how clinicians go about making diagnoses by seeing patient histories through their 'trauma lenses'. If trauma-informed assessments became the norm in the mental health field, it would dramatically shift how diagnosis is thought of as well as how individuals are understood and treated. (Harris & Fallot, 2001). As it stands, most people in services who have experienced trauma are not being treated for the effects of said trauma. Often times their clinicians are not even aware of the traumatizing events (Harris & Fallot, 2001). This could be harmful to the client's recovery by ignoring the root of the problem and potentially re-traumatizing the client.

One line of evidence that supports the idea that trauma more adequately explains the symptoms of bipolar disorder than an inherent mental illness is the similarities between the diagnostic criteria listed in the DSM and the effects of trauma that have been laid out by trauma theory. The defining characteristic of any bipolar diagnosis is an episode of extreme mood; whether it be mania, hypomania or depression (American Psychiatric Association, 2013). People who have been traumatized often have all or nothing thinking, meaning they are extremely activated or low in mood. Just like people with a diagnosis of bipolar disorder, traumatized people have difficulty matching their emotional reaction of something to the real severity of that situation (American Psychiatric Association, 2013, Bloom, 1999). Other parallels between trauma and the DSM description of bipolar disorder include irritability, which is associated with both a manic episode, as well as the effects of a traumatized psyche. Impulsivity, a hallmark of mania, is also associated with trauma theory, because as described above, people who have constantly been in fear for their life tend to react on instinct.

An understanding of trauma reenactment could also help explain the typical presentation of someone who is diagnosed with bipolar disorder. People who have been traumatized are subconsciously attempting to re-feel their trauma while at the same time doing anything to escape the anxiety they are experiencing. People with bipolar disorder tend to have unstable relationships; a trauma lens could help us see the underlying mechanism that causes distress in the individuals' interpersonal lives.

The rates of childhood trauma in people already diagnosed with bipolar disorder not only explain these similarities but also support the shift to trauma-informed assessment in mental health practice. Many different studies (Etain et al. 2008, Larsson et al. 2013, McDonnell, 2010) show a link between a diagnosis of bipolar disorder and adverse childhood events and/or conditions. These findings suggest more than just a genetic link and suggest a causal effect of traumatic experiences.

While not every person who is diagnosed wth bipolar disorder has experienced trauma as describe by trauma theory (violent crimes, combat, etc.) the understanding of what constitutes trauma is expanding (Center For Nonviolence And Social Justice, 2014). If we go back to our definition of trauma "when inner and outer resources are inadequate to deal with external stress" then many more adverse conditions qualify. Such conditions could include racial discrimination, poverty, community violence and other issues experienced in urban areas (Center For Nonviolence And Social Justice, 2014). Research shows that the number one demographic most diagnosed with bipolar disorder is young low-income African American males (McDonnell, 2010). Given the link between such traumatic events and manifestation of DSM criteria for diagnosis, this is not surprising.

Although genes play a role in how trauma will be experienced, there has been no specific gene found that supports a neurobiological etiology of bipolar disorder. Rather it is more likely that some people, for a number of different reasons--including genetic factors, are more vulnerable to psychological distress when put through traumatizing conditions. The unfortunate reality is that most of these people will receive a mental diagnosis and be treated based on that diagnosis rather than being able to explore and understand their trauma and overcome the effects it has had on their psychological functioning (Harris & Fallot, 2001).

While diagnosis may serve as a support for some people in mental health services, by providing a seemingly objective explanation for the way they feel and how they act, for many, especially those with trauma backgrounds, diagnosis can be very disempowering. The good news about trauma is the effects can be unlearned through long-term psychotherapy; a DSM diagnosis of bipolar disorder implies that these expressions are inherent and cannot be changed.

Trauma theory also explains how in traumatic conditions individuals often feel helpless, and over time they begin to internalize this helplessness; this is referred to as 'learned helplessness' (Bloom, 1999). Because of the subjective experience of learned helplessness, people who have withstood traumatic experiences would benefit from a more nuanced clinical understanding: an understanding that these feelings and experiences of irritability, impulsivity, anxiety and/or dissociation are not a problem but just learned reactions based on what they have endured in their lives.

Coping skills that are necessary for survival under conditions of traumatic stress become learned over time. When faced with life-threating trauma, the ability to act on impulse and numb emotions may be beneficial, even critical to survival. Akiskal and Hagop (2005) suggest that there are evolutionary benefits to the clinical presentation of bipolar and other affective disorders. It goes to show that such expressions have served the traumatized person in times of crisis. It is not until the experience of the trauma, and these survival instincts become internalized over time that they no longer are adequate for emotional protection and lead instead to emotional distress.

Conclusion

A review of the literature on bipolar disorder and childhood trauma can help clinicians understand the importance of trauma informed assessment. When we experience overwhelming emotional conditions we are unable to cope, our selves become fragmented and our psyches become traumatized. It is less helpful to give a mental diagnosis, that suggests genetic deficiencies, than to take into account an individual's history that potentially has traumatized them, leading to the present symptoms. A major paradigm shift in the field of mental health would be required if trauma-informed assessments were to become used regularly in practice. However it would greatly benefit consumers if we took to heart this more nuanced understanding of the human person.

References

Akiskal, K. & Hagop, S. (2005) Theoretical Underpinnings of Affective Temperaments: Implications for Evolutionary Foundations of Bipolar Disorder and Human Nature. *Journal of Affective Disorders.*

American Psychiatric Association. (2013). Bipolar and Related Disorders. In *Diagnostic and Statistical manual of Mental Disorders* (5th ed.)

Bloom, Sandra (1999) Trauma Theory Abbreviated. In *The Final Action Plan.* Attorney General of Pennsylvania's Family Violence Task Force. Pp. 2-13

Center for Nonviolence and Social Justice (2014). What is Trauma? Trauma Theory and the Effects of Trauma. *PeacePoints.*

Etain, B. et al. (2008). Beyond Genetics: Childhood Affective Trauma in Bipolar Disorder. *Bipolar Disorders.* Vol. 10. Pp. 867-876.

Geller B. & Luby J. (1997). Childhood and Adolescent Bipolar Disorder: A Review of the Past 10 Years. *Journal of American Child and Adolescent Psychiatry.* 39:9. Pp. 1168-1176.

Harris M. & Fallot R. (2001) Envisioning a Trauma-Informed Service System: A Vital Paradigm Shift. *New Directions for Mental Health Services.* Vol. 2001: Issue. 89. Pp. 3-22.

Larsson, S. et al. (2013). Patterns of Childhood Adverse Events are Associated with Clinical Characteristics of Bipolar Disorder. *BMC Psychiatry.* 13:97.

McDonnell M. (2010). Race, Gender and Age effects on the Assessment of Bipolar Disorder in Youth. *Northeastern University Nursing Dissertations.*

Asperger's Syndrome:
Disorder or Personality Trait

Suzanne Wychocki

A Brief History of Asperger's Syndrome

As early as the 1300's in Jean Itards's subject Victor, the wild boy of Aveyron, there has been the prevalence and recognition of autistic behaviors in society (Tsai & Ghaziuddin, 2013). Despite the historical observation of autism, it was not until the early 1900's that the disorder was actively diagnosed. The word *autism* stems from the Greek word 'autos' meaning 'self' and 'ismos' meaning state or condition; isolate or self-condition. In 1943 the physician Leo Kanner associated the term autism to describe a group of eleven children with 'deficits in their ability to relate to other people' and noticeable restrictions in certain behaviors. (Tsai & Ghaziuddin, 2013) The term autistic psychopathy was originally used in 1911 by the Swiss psychiatrist, Eugen Bleuler to describe a group of schizophrenic symptoms. (Tsai & Ghaziuddin, 2013) The following year, Viennese psychiatrist Hans Asperger released his study of four boys who showed marked 'autistic withdrawal,' the symptom, at the time still associated with schizophrenics (per Bleuler's use of the term several decades earlier; Tsai & Ghaziuddin, 2013) Thus the term *autistic* came to be associated with the diagnosis of schizophrenia and forms of psychosis. This was never Asperger's intent, but nevertheless it created decades of confusion around the actual definition of the term and what constituted the criteria for each diagnosis.

Bridging the language and geographical divide with other developing countries and studies did not occur until the early 1980's when Lorna Wing, a British psychiatrist, produced her own research on the diagnosis of autistic tendencies. She chose to refer to the cases as Asperger's Syndrome rather than autistic psychopathy as the professional inclination at the time was to correlate this with psychopathy and therefore with sociopathic behavior which was not inherent in her patients. (Molloy & Vasil, 2002)

Introduction into the DSM

Appearing in the DSM-III in 1980 under the name Infantile Autism and classified under the category of pervasive developmental disorders (PDDs), emerged the newly coined Autism Spectrum Disorder. There was no mention of Asperger's Syndrome, as it was not yet identified in the English language. Wing's paper in 1981 began to shed light on what she considered the 'autistic continuum.' (Molloy & Vasil, 2002) In the thirty years since it's formal introduction into the world of mental health there have been many iterations of Asperger's Syndrome. With noticeable and possibly questionable indecisiveness of its formal name, Asperger's Syndrome has been know as Asperger's Disorder, High Functioning Autism, or mild Autism to name just a few. The added significance to the diagnosis is its obvious inclusion into the DSM-IV under the PDD category and then its removal and complete elimination from the DSM-V ten years later. What have not been eliminated are the individuals who have been diagnosed with Asperger's Syndrome during the time the diagnosis appeared in the DSM. Does this mean that these individuals no longer have a disorder or could it be possible that they never had a disorder to begin with and were labeled under certain mental health guidelines and pressure within the societal constructs of the public, the education system and the society at large? In this paper I will explore the social constructionism of Asperger's Syndrome (now called Autism Spectrum Disorder Level 1 in the DSM-V which I will continue to refer to as Asperger's Syndrome as many clinicians and patients continue to do) and the benefit of seeing it as a personality trait that should be celebrated in our society rather than a disability caused by a biological component.

Asperger's Syndrome Traits

Criteria for Asperger's Syndrome diagnosis using the DSM-IV would have been a presence of qualitative impairment in social interaction and restrictive, repetitive and stereotyped patterns of behavior, interests and activities. These patterns must have been substantial enough to interfere with normative social functioning. (Kite, Gullifer, Tyson, 2012) The differentiation between Autistic Disorder and Asperger's Syndrome was predicated on the level of cognitive development of the individual. Asperger's Syndrome patients showed no delays in cognitive development while Autistic Disorder patients had signs of cognitive delay.

Asperger traits can range from limited ability to play appropriately with other age-related children in a school or playground setting, lack of ability to join into ongoing conversations with the expected input and intense interests in topics that distract from normal play or behavior. (Young & Rodi, 2013) Each of these traits is based on a predetermined social construct of what is considered normal for the environment that the child is currently exposed to. Who decides what is normal? The issue becomes less about the child's behaviors and more about what the adults around them constitute as normal and what they are able to tolerate.

The recent increase in the diagnoses of autism spectrum disorders and specifically Asperger's Syndrome can be seen to correlate to the need to assign a diagnosis or label to a student as young as three years old in order for the schools and the teachers educating those students to receive additional funding for services. The emphasis of these interventions is 'normalizing' the child. The services provided for these students range from social norms groups, extended time and additional accommodations for testing and individual therapy. While there are also resources made available to the students themselves to help with their social emotional development through one-on-one counseling within the school, most of the resources are used to improve the grades and standardized test scores for these students and to help normalize them to the social norms of the public school. Most are in agreement that individual counseling is a positive benefit of interventions for a student with Asperger's Syndrome, but it also raises the question of offering individual counseling for all students within the school system as each child is navigating his or her own struggles though adolescence and young adulthood. For example, a child with an extreme interest in airplanes (or whales, spiders or any other subject not covered in standardized schooling) may spend an inordinate amount of time focusing on that subject. It will depend on that child's family, school and community if this interest becomes a passion or a symptom of a disorder. There is also the view from those diagnosed with Asperger's Syndrome that their lack of social interaction is not necessarily due to not knowing how to respond in certain situation but a conscious decision not to engage with others in 'small talk' that they personally find trivial. Adults acting in this manner may be considered aloof but children exhibiting this behavior are labeled as lacking in empathy and considered deficient because they do not fit into our societal norms of conversational banter. (Rogers, et al, 2006)

Updates to the DSM Criteria

As with many diagnoses in the DSM, the level of variables and probability of diagnosis is based solely on the clinician's interpretation of the individual, the current environment and the clinician's tolerance for differences in social norms. In most western cultures, where Asperger's Syndrome is most prevalent, (as well as the diagnosis of ADD and ADHD) there has been a marked increase in the frequency of diagnosis. Thirty years ago the prevalence of Autism Spectrum Disorders was estimated at 1 in 2,000 in the United States. The inclusion of this new category in the DSM-IV in 1994 correlated to an increase in diagnosis for ASD of 600%. In 2007 that number had increased to 1 in 155 then to 1 in 88 in 2012. (Fien, M, 2015).

Prior to the release of the DSM-V, Asperger's Syndrome was delineated by decreased and odd social interactions and restricted interests and behaviors. The removal of Asperger's Syndrome from the DSM-V forced clinicians to alter the diagnosis of well-established individuals with Asperger's Syndrome under the DSM-IV to the new umbrella category Autism Spectrum Disorder Level 1. The caveat being that those already carrying the label of Asperger's Syndrome will have to either be reevaluated or automatically given a new diagnosis, thus altering the services provided to them. The direct connection between the DSM-V, the medical fields and insurance companies determines which services are retained and which are removed.

Those individuals not meeting the new criteria for true autism spectrum disorder (this may also include High Functioning Autistics, a group with similar traits to Asperger's Syndrome, as there is little differentiation made between Asperger's and High Functioning Autistics and many consider them the same disorder) would be diagnosed with Social (Pragmatic) Communication Disorder 315.39 (F80.89). The diagnostic criteria for Social (Pragmatic) Communication Disorder are on par with the communication criteria for ASD Level 1 minus the symptoms for repetitive behaviors. (APA, 2013)

There is limited data as to the prevalence of Asperger's Syndrome within the data for Autism Spectrum Disorder. There have been estimates that only 80% of those diagnosed with Autism Spectrum Disorder under the criteria for Autism Disorder in the DSM-IV retain the diagnosis under the criteria for the DSM-V. (Young & Rodi, 2013) The remaining 20% either miss the diagnosis by not meeting all of the criteria or will fit into another diagnostic category. (Young & Rodi, 2013) This in and of itself begs the question 'What about the 20%?' Were these children diagnosed because their communication styles and social desires were not considered normal or do they really still need a clinical diagnosis and interventions within the school?

There are many theories as to the increase in ASD diagnosis, including the introduction of the category in the DSM, which made diagnosis more accessible and the increase in media awareness of the subject. There have also been changes in the public school systems as to how special services are allocated to children. As teaching models become more rigid, and standardized testing increases, teachers job evaluations are affected and dissemination of monies across school districts is impacted, and these intercorrelations may contribute to the increase in Asperger's Syndrome diagnosis. More research on this is needed but the increase in Asperger's Syndrome diagnosis (as well as ADD and ADHD) may have a greater correlation to the increase in demands on school children in western cultures.

By creating the label (diagnosis) of Asperger's Syndrome in the 1980's our clinical professionals created a box to check for those students who did not behave in accordance with the current standards of the public school system. As most children were not diagnosed with Asperger's Syndrome until they entered the structured school system, it wasn't until they were placed side-by-side with other children at the same chronological age that some of their behaviors began to seem different. It is accepted in the development of every child that they are individual beings and will grow and develop at different rates. While it is helpful to create guidelines and expectations for parents to track the development of their children, it is perfectly acceptable to say that every child is different and will develop at a different rate. Simply because our school system is structured around the premise that groups of students should be put together because they happen to be born within a similar timeframe does not guarantee that they will develop at the same rate emotionally or physically. Intuitively we all know this, and if we look at a classroom of first graders we would not expect them to be developing at the same rate. Knowing this does not change the fact that our schools are still set up this way, and we continue to use benchmarks and test scores to evaluate the success of our children. This also applies to the social development of all children. A classroom of 30 students will see a wide range or a spectrum of social skills and development at any given time during the school year.

The structure of the school day, while sounding like the perfect antidote to the inflexible behaviors of a severely autistic child, becomes challenging and uncomfortable for a child who may be diagnosed with Asperger's Syndrome. Other children who may not already have a diagnosis or are quite possibly just the wrong fit for the environment may also be stressed by the structure of the school day. The rigidity of the public school classroom is replicated in no other environments in our society except possibly its predecessor from the industrial revolution, the factory setting. Prior to entering the school system most children spend their days freely and with little structure except meals and sleeping patterns. After they leave the school system and enter into adulthood, they are able to return to a system and structure more of their choosing that fits their personality traits. This applies to all children including neurotypical children as well as children who may suffer from an extremely mild case of Asperger's Syndrome. Is it possible that at some level all humans suffer from a form of Asperger' Syndrome or fall somewhere on the Autism Spectrum?

Another Scenario

Imagine the following scenario: Compare the level of intensity of the symptoms of Asperger's Syndrome with the symptoms of a common condition seen in many social environments, *Lack of Self-Esteem*. You could possibly see some similarities in their effect on the child and their personal development as well as socialization within the classroom. While *Lack of Self-Esteem*, is not a true disorder according to the DSM-V, the behaviors associated with it could also be considered abnormal. Symptoms of *Lack of Self-Esteem* can have a similar effect on the social interactions of school-age children as well as adults. Individuals who show signs of *Lack of Self-*

Esteem do not take criticism or constructive feedback well. They may be overly critical of others rather than being supportive. Rather than focusing on personal growth they spend time worried about not making mistakes, and they are prone to bullying behaviors that can be detrimental to other classmates and the culture within the school.

Now, I realize that this is an exaggeration, but for the purpose of making my point the current public and clinical views of Asperger's and our tendency to view thesebehaviors as symptoms of a disorder that need to be corrected could really be unique expressions of their personality traits that should be accepted by our 'normal' society. School children with *Lack of Self-Esteem,* are often considered bullies, have a tendency to cheat more, use avoidance as a coping mechanism and can also be more prone to peer-pressure as they get older. All of ese are negative and abnormal behaviors in an optimal setting, but have been socialized to be more acceptable behaviors in our aggressive, independent society. (Rogers, et al, 2006) These are viewed as pro-social traits whereas intensity, shyness and focus, when seen together as flaws, create the need for a medical diagnosis. None of these behaviors are addressed with the special interventions that Asperger's Syndrome are addressed unless there is comorbidity with another disorder or learning disability.

Viewing the public school classroom or even a typical workplace mixed with varieties of individuals that represent both the autism spectrum with their diagnosed limited social skills and those exhibiting traits of *Lack of Self Esteem*, it is hard to discern which disorder is more in need of intervention. A compassionate teacher would treat both children with the same kindness but realize that they are on varying ends of the social-emotional spectrum and need to be handled differently. Is it possible that increasing the focus on social-emotional learning in the classroom and teaching diversity beyond race and gender would be a benefit to students and teachers no matter where they fall on the spectrum?

Aging out of the Asperger's Diagnosis

As an individual ages and moves through the school system there have been many opportunities to increase their chances of having a successful and satisfying life. Now, this alone is a subjective decision. What determines for others what a successful and satisfying life really is? Only the mind of that person can decide that. The restrictive, repetitive behaviors that may not have been appropriate in the classroom where many topics are covered everyday, becomes an asset to an adult and potential job seeker who needs the skills of a focused, deeply knowledgeable employee. A fascination with maps and geography will be extremely beneficial for an aspiring cartographer.

On the other hand, an individual suffering from *Lack of Self-Esteem* that is not addressed with the same vigor and methodology as an Asperger's student will continue to experience a poor quality of life, engage in potentially harmful disrespectful behaviors towards others and risk an increase in anxiety.

As the populations of individuals who have been diagnosed with Asperger's Syndrome while it appeared in the DSM-IV have grown up, they have become more vocal about their diagnosis. Being able to put a label on one's traits can be helpful in some cases the same way a person with moderate Obsessive Compulsive Disorder (OCD) may take pride in their orderly lives, a person with Asperger's Syndrome may see their intense interest in a certain subject and focus as a positive attribute of their lives. As adults we have all come to recognize the famous and successful people in our world who are said to have Asperger's Syndrome (formally diagnosed or not) that have made great strides in our culture and society.

Summary

It is obvious from my research that the discussion surrounding Asperger's Syndrome and Autism Spectrum Disorders needs to continue in conjunction with increased research on brain development, social-emotional learning and communication styles. As there is no clear set of diagnoses from a strictly medical perspective, it can be said that any clinician will use her own experiences, judgments and personal tolerances to diagnosis a child with Asperger's Syndrome. This may not be a decision that can be left solely on the shoulders of a clinician but needs to be looked at in conjunction with educational policies, public discourse on 'normal' behavior and an increase in empathy for varying personality traits.

References

American Psychiatric Association., & American Psychiatric Association. (2013). *Diagnostic and statistical manual of mental disorders: DSM-5*. Washington, D.C: American Psychiatric Association.

Chew, B., & Rosen, L. (2008). Autism/asperger's syndrome. In F. Leong (Ed.), *Encyclopedia of counseling*. (Vol. 1, pp. 35-37). Thousand Oaks, CA: SAGE Publications, Inc. doi:

Duffy, Frank H, Shankardass, Aditi, McAnulty, Gloria B, & Als, Heidelise. (2013). *The relationship of Asperger's syndrome to autism: a preliminary EEG coherence study*. (BioMed Central Ltd.) BioMed Central Ltd.

Fein, E. (March 01, 2015). "No One Has to Be Your Friend": Asperger's Syndrome and the Vicious Cycle of Social Disorder in Late Modern Identity Markets. *Ethos, 43,* 1, 82-107

Ghaziuddin, M. (January 01, 2011). Asperger disorder in the DSM-V: sacrificing utility for validity. *Journal of the American Academy of Child and Adolescent Psychiatry, 50,* 2, 192-3.

Kite, D., Gullifer, J., & Tyson, G. (2013). Views on the Diagnostic Labels of Autism and Asperger's Disorder and the Proposed Changes in the DSM. *Journal Of Autism & Developmental Disorders, 43*(7), 1692-1700. doi:10.1007/s10803-012-1718-2

Ozonoff, S. (2012), Editorial Perspective: Autism Spectrum Disorders in DSM-5 – An historical perspective and the need for change. Journal of Child Psychology and Psychiatry, 53: 1092–1094. doi: 10.1111/j.1469-7610.2012.02614.x

Rogers, K., Dziobek, I., Hassenstab, J., Wolf, O. T., & Convit, A. (January 01, 2007). Who cares? Revisiting empathy in Asperger syndrome. *Journal of Autism and Developmental Disorders, 37,* 4, 709-15.

Tsai, L. Y., & Ghaziuddin, M. (February 01, 2014). DSM-5 ASD Moves Forward into the Past. *Journal of Autism and Developmental Disorders, 44,* 2, 321-330.

Young, R., & Rodi, M. (2014). Redefining Autism Spectrum Disorder Using DSM-5: The Implications of the Proposed DSM-5 Criteria for Autism Spectrum Disorders. *Journal Of Autism & Developmental Disorders, 44*(4), 758-765. doi:10.1007/s10803-013-1927-3

Gender Inequalities Among People Living with Generalized Anxiety Disorder

McKenna Lynch

In the Fifth Edition of the *Diagnostic and Statistical Manual of Mental Disorders* (DSM-5), there are twelve Anxiety-Related Disorders. According to Craske et al (2003), an Anxiety Disorder is a future-oriented state of mind associated with consistent worry and fear of negative events. While Anxiety Disorders tend to be co-morbid, the disorders are vastly different. An Anxiety Disorder is classified by the thoughts or beliefs one has, in addition to the type of situation an individual fears or avoids (DSM-5. 2013). Anxiety is a treatable condition, and an individual has potential to live a high quality life. However only one-third of individuals with an Anxiety Disorder receive treatment.

It is important to differentiate between natural anxiety and an Anxiety Disorder. It is healthy for humans to feel anxious during difficult situations; for instance a job interview or moving to a new town. When an individual worries excessively and fears seem overwhelming, the individual may have an Anxiety Disorder. Summarized in the criteria of the DSM-5 (2013) an anxiety disorder is when the worries and fears interfere with one's daily life. Anxiety-Related Disorders are prominent and affect 18% of adults in the United States, which are roughly forty million individuals. Compared to men, women are 60 % more likely to develop an Anxiety Disorder within their lifetimes (National Institute of Mental Health).

Diagnosis

Generalized Anxiety Disorder (GAD), is a disorder that is characterized by the presence of excessive anxiety and worry. When the worrying is uncontrollable it becomes a presenting concern, which often leads an individual to seek out professional help (DSM-5, 2013). An individual who lives with GAD encounters severe anxiety and worry, along with mental and physical symptoms. According to the DSM-5 (2013), within a 12-month range, GAD affects approximately .09% of adolescents and 2.9% of adults in the United States. The disorders that tend to be co-morbid with Anxiety Disorders are Depression Disorders, Eating Disorders, and Attention Deficiency Hyper Disorder (DSM-5, 2013).

There are three main risk factor categories that are believed to influence the onset of GAD; temperamental, environmental, and genetic and biological. Temperamental factors are negative affectivity, behavioral inhibition, and harm avoidance. Childhood adversities and parental overprotection would be categorized under environmental factors. One-third of the risk of experiencing GAD is genetic (DSM-5, 2013).

Hettema et al. (2001) reviewed similar literature and previous studies on GAD within families. The studies suggest that GAD is likely due to the genetic effect of the familial aggregation. There is no one definite cause of anxiety, and a clinician and client should be educated on that. As seen above there can be a variety of risk factors that may make an individual more susceptible or vulnerable to GAD. This critique specifically examines the many risk factors that generally apply to women.

The DSM provides a guideline of what clinicians should be aware of in clients and patients. The DSM-5 (2013) describes symptoms of Generalized Anxiety Disorder as including restlessness or feeling on edge, fatigue, difficultly concentrating, irritability, muscle tension, and/or sleep disturbance (DSM-5, 2013). An individual must experience three of the above symptoms for more days than not, lasting for six months. Physical symptoms and anxiety can cause distress and impairment in one's collegial life at work, in social situations and other areas of daily life. (DSM-5, 2013). The DSM-5 does not list physical symptoms, but according to The National Institute of Mental Health (2015), physical symptoms may include fatigue, headaches, sweating, and nausea.

Women and Generalized Anxiety Disorder

Men and women express anxiety symptomatology differently. This analysis of Generalized Anxiety Disorder will discuss women who live with GAD, briefly highlighting the gender differences among men and woman. Various models offer different viewpoints of GAD, helping explain gender discrepancies. The two theories that will be used to elevate the understanding of GAD are Objectification Theory and Attachment Theory.

Goldenberg and Roberts (2004) argue the common societal discrimination women experience. Another viewpoint of Objectification Theory is from Fredickson and Roberts (1997-06), which discusses the consequences of being a woman in culture that promotes the objectification of the female body. Cassidy et al. (2009), states that the Attachment Theory supports the relationships an individual may have in childhood and how those relationships can have implications to anxiety later on in ones life. These theories will provide a framework for interpreting gender variance among Generalized Anxiety Disorder.

Women are twice as likely to experience symptoms of GAD as men (DSM-5, 2013). Individuals living with GAD may even start to experience excessive worrying as early as childhood. Many individuals who have GAD have felt anxious and worried the majority of their lives (Cassidy et al 2009). One's likelihood of diagnosis of GAD poses a question - why are woman being diagnosed twice as much as men?

Walf and Frye (2006), state that estrogen can affect serotonin levels in women's brains, which affects the central nervous system. The researchers found that women are more vulnerable to mood disorders post puberty, such as GAD. Levels of estrogen are significantly lower among women with mood disorders and may increase the symptomology (Walf and Frye, 2006). Mufson (2006), examines the hormonal and biological factors of GAD. The text describes the relationship estrogen has in regulating a women's mood, sleep, and appetite. Considering this information, some may infer that women are more likely to encounter anxiety.

There are several other factors that researchers believe contribute to high rates of GAD in women. According to the National Domestic Violence Hotline, nearly three in ten women have experienced rape, physical violence and/or stalking by a partner. The women reported that the traumatic experience impacted their functioning. Craske (2003), states that when a woman experiences a major traumatic event, she is more prone to have emotional distress. Musfson (2006), agrees with the statistics of abuse and discusses the relationship between abuse and the long-term harm that can change the brain and brain structure. This relationship may make women more vulnerable and could be a predisposition to form an anxiety disorder due to the increased probability of encountering a traumatic event, such as rape or physical violence.

Craske(2003), explains that it is culturally accepted that many women take the primary responsibly as a caretaker within the family. These responsibilities can cause a great amount of stress for a woman. Evolutionary factors, such as women taking care of offspring, can increase a women's fight or flight responses, causing her to be more anxious and protective. Women described being aware of their states of mind more than men. Internal awareness about one's self can create fear and anxiety in women because one is conscious of most of one's thoughts, feelings, and behaviors. Worrying can interfere with daily activities, cloud problem solving, and contribute to the cycle of anxiety (Craske, 2003).

Vogel et al. (2007), state when seeking out help, women tend to have more positive attitudes compared to men. This finding may impact the reason why women are diagnosed and receive treatment more than men. When women are diagnosed with less severe diagnoses, such as Generalized Anxiety Disorder, women continue to seek out more help then men. Women who are seeking out help more frequently are at a higher chance of being diagnosed with another mental illness such as depression, which is often co-morbid with anxiety disorders (Vogel, 2007). To better examine GAD, Objectification Theory and Attachment Theory will provide a gender inequality perspective.

Objectification Theory and Attachment Theory

Objectification Theory posits there are societal barricades that uniquely affect women who are exposed to gender inequalities in the United States. For example, objectification is integrated throughout social norms within the United States through the media, school systems, pay grades, and women's personal experiences. Women and young girls are faced with great pressure to appear a certain way, increasing consciousness of their physical appearance (Golberg and Roberts, 2004).

The objectification tends to involve societal gender roles and discrimination directed at women. Objectification can be associated with high psychological consequences. Golberg and Roberts (2004) state that women are often devalued in specific qualities that are not often characterized by men, such as being more emotional.

Fredickson and Roberts (1997-06), view Objectification Theory as an agent to help understand the serious cost women encounter by being in a society that sexualizes the female body. Internalizing physical appearances can increase the likelihood of a woman experiencing shame and anxiety. Women tend to internalize the appearance of attractive women in society and the perceived power that is coupled with it. The study suggests that a disturbed conscious results in a woman internalizing the practices of a culture's objectification in combination with constantly monitoring her appearance. These occurrences influence one's psychological well-being (Fredickson and Roberts, 1997-06).

An individual experiences anxiety when he or she perceives a threat. Culture tends to promote objectification of females, therefore, leading women to continually experience anxiety-provoking events and situations because she feels threatened. Fredickson and Roberts (1997-06), discuss how this overwhelming worry and anxiety maintain a chronic awareness of women's physical appearances and her overall physical safety.

Cranske (2003) also discusses that females react more to facial expression, which is related to one being fearful or anxious as well as communicating threat. Cranske (2003) states, " in general, by virtue of elevated negative affectivity, females are at greater risk for detecting threats, learning threat associations, and retaining high levels of reactivity to aversive events" (p.200). It can be physically and emotionally taxing to recognize and be observant to the threats that are present, which can cause one to worry excessively.

Objectification Theory examines gender specific risk factors that may increase a women's likelihood to be diagnosed with GAD due to cultural norms. There are several other risk factors that influence the onset of GAD. These factors include family history of anxiety, existing mental illness, low socioeconomical status, and history of physical or emotional trauma. Trauma can include childhood separation or over-protective child rearing (Cranske, 2003). Attachment Theory attempts to explain one's interpersonal relationships with others, for instance family members or the primary caregiver.

Bowlby's Attachment Theory presents a framework for development of Anxiety Disorders (Cassidy, 2009). Bowlby states that adults with an Anxiety Disorder may have been exposed to vulnerabilities in childhood. Poor attachment experiences may leave the child feeling uncertain when he or she experiences a time of distress or trouble. Clients who were diagnosed with severe GAD reported experiencing less maternal love in childhood, greater maternal rejection/neglect, more maternal role-reversal, and enmeshment compared to the controlled participants. Clients living with GAD also reported currently having a greater vulnerability in regards to their mothers and find it challenging to access childhood memories (Cassidy, 2009)

According to Cassidy et al. (2009), the Dysregulation Model of GAD developed by Mennin and colleagues (2000), enriches the avoidance model. This framework explains why individuals living with GAD tend to avoid emotions, whether positive or negative. Individuals with GAD experience intense negative emotions, making it difficult to identify and regulate their emotions. When emotions are heightened, individuals with GAD seek out emotion-regulation strategies, such as worrying to avoid the onset of impending suffering. Research has found that poor interpersonal relationships can contribute to the development and maintenance of GAD. Individuals living with GAD reported seeing themselves as interpersonally less effective than others and are believed to be overly nurturing and intrusive in relationships. There is a strong association between GAD and maternal difficulties, lack of friendships, and low family cohesion (Cassidy et al, 2009).

Recently, research, theory, and treatment have expanded the predisposing factors that may influence diagnosis of GAD and have increased focus on interpersonal issues. Parts of emotion regulation are related to attachment factors, such as positive or negative childhood attachments. The research conducted by Cassidy and colleagues (2009) provides a more complex understanding of the factors that affect the development of GAD and how current attachments and relationships contribute to the development and maintenance of GAD (Cassidy et al, 2009).

Men and Generalized Anxiety Disorder

From a young age, even just a few months old, social patterns of reinforcement between boys and girls are drastically different. Parenting differences foster certain abilities in children, for example the development of one's strengths or weaknesses. Craske (2003), found that mothers show more meaningful connections with their sons than daughters. According to Carske (2003), one can speculate that young boys have intense emotional needs. Meaningful connections are vital in one's development because they determine a bias for "…predictability and controllability over the world and over one's own emotional responding as well as tensional control, which in turn [is] critical to levels of ongoing distress and ability to regulate emotional arousal" (Craske

2003, p. 200). The importance of healthy attachments creates positive emotional regulation, and one may be less likely to develop GAD.

During childhood there are reinforced behaviors, such as boys are to be more aggressive, physical, and independent. Girls are encouraged to control their behaviors and anxious behaviors are reinforced. The gender differences support the idea that boys are more capable than girls in childhood, which is transformed into adolescence and adulthood (Craske, 2003). However, one can also argue that the pressure young boys have to be independent and in control could also cause worrying, fear, and anxiety.

Although there is a higher prevalence rate of women being diagnosed with GAD in their lifetimes, men also experience chronic anxiety. Men are often not treated for Anxiety Disorders and may overlook their symptoms. For example, when men experience a panic attack, professionals may misdiagnosis it as a heart attack. Anxiety Disorder statistics may not accurately reflect the number of males with this diagnosis due to under reporting. Overlooking or under reporting symptoms can lead to misdiagnosis or under diagnosis in men (Craske, 2003).

The same cultural biases that influence women's diagnosis with GAD also influence men not receiving the diagnosis. Biases can include stigma and shame, which influence one seeking out treatment. According to Bolton et al (2006), it is generally not accepted for a man to admit he is enduring pain or is experiencing anxiety. Men are more likely to self-medicate if they are experiencing bouts of anxiety. Common self-medication includes alcohol or drugs, which help them relax. The self-medication can also contribute to anxiety, feeding a cycle of dangerous coping strategies (Bolton et al, 2006).

Discussion

Objectification Theory and Attachment Theory both provide a strong framework when looking at the gender inequalities in GAD. When considering Objectification Theory it easily lays out how culture and society is deeply integrated among women objectification. Objectification Theory explains how internalizing body image paranoia and safety increases one's anxiety. Attachment theory examines early attachments and how poor attachments can cause one to feel unsure and anxious in troubling situations. Uncertainties in childhood can turn into an Anxiety Disorder as one gets older.

There has not been solely one factor that contributes to GAD, but a wide variety. As discussed previously, culture, genetics, biology, and one's upbringing have a large impact on one's vulnerability to development of GAD. Objec-Objectification Theory and Attachment Theory both discuss the inequalities and differences women experience compared to men, in regards to the likelihood of diagnosis of GAD. Knowing the factors and symptoms of Generalized Anxiety Disorder will hopefully create awareness and equality among men and women in the United States regarding all aspects of ones life; that would be a huge contribution to Generalized Anxiety Disorder.

References

Behar, E., DiMarco, I.D., Hekler, E.B., Mohlman, J., & Staples, A. (2009). Current theoretical models of generalized anxiety disorder (GAD): Conceptual review and treatment implications. Journal of Anxiety Disorders, 23, 1011-1023. doi:10.1016/j.janxdis.2009.07.006

Bolton, J. (2006-11). Use of alcohol and drugs to self-medicate anxiety disorders in a nationally representative sample. The journal of nervous and mental disease, 194(11), 818-825.doi:10.1097/01.nmd.0000244481.63148.98

Cassidy, J. (2009-03). Generalized Anxiety Disorder: Connections With Self-Reported Attachment. Behavior therapy, 40(1), 23-38.doi:10.1016/j.beth.2007.12.004

Craske, M. G. (2003). Origins of phobias and anxiety disorders: Why more women than men? Oxford, UK: Elsevier.

Fredrickson, B. L., and Roberts, T.A. (1997-06). OBJECTIFICATION THEORY. Toward Understanding Women's Lived Experiences and Mental Health Risks. Psychology of women quarterly, 21(2), 173-206.doi:10.1111/j.1471-6402.1997.tb00108.x

Goldenberg, J. L., and Roberts, T.A. (2004). The beast within the beauty: An existential perspective on the objectification and condemnation of women. In: J Greenberg, SL Koole, T Pyszczynski (eds), Handbook of experimental existential psychology; p 71–85. (Guilford Press).

Mufson, M. (2006). Coping with Anxiety and Phobial. Harvard Health Publications.

National Institute of Mental Health. Anxiety Disorders. Retrieved April 6, 2015, from http://www.nimh.nih.gov/health/topics/anxiety- disorders/index.shtml#part_145339

The National Domestic Violence Hotline. Statistics. (n.d.). Retrieved March 29, 2015, from http://www.thehotline.org/resources/statistics/

Vogel, D. L. (2007-10). Avoidance of Counseling: Psychological Factors That Inhibit Seeking Help. Journal of counseling and development, 85(4), 410-422.doi:10.1002/j.1556-6678.2007.tb00609.x

Diagnosing Substance Use Disorders in Adolescents Using Attachment Theory

Melissa McMahon

Substance use disorders are among the many categories changed with the fifth of edition of the *Diagnostic and Statistical Manual of Mental Health Disorders* (DSM-V). The preface of the DSM-V (2013) states that this category was altered with the, "hope to alleviate some of the widespread misunderstandings about these issues," (p. xlii). The altering of this diagnostic category has shed new light on the issue of substance use and has redefined the diagnosis of substance use disorder. The DSM-V now defines substance use disorders as using an excessive amount of alcohol, caffeine, cannabis, hallucinogens, inhalants, opioids, sedatives, hypnotics, tobacco, and/or other substances. These drugs are taken in a variety of ways and activate the brain's reward system, making the user have a positive biological reaction to the intake of these substances (American Psychiatric Association, 2013).

Those who fall under the diagnostic category of substance use disorders experience several primary criteria, according to the DSM-V, that identify their disorder. The first of these is continued use of a substance despite substance-related problems. A simple example of this would be continuing to binge drink alcoholic beverages despite experiencing symptoms of a hangover the next day. The next criterion is social impairment. This implies that the use of a substance is in some way preventing the user from having success socially in work, school, home, or other obligations. Another criterion for substance use disorder is risky substance abuse, which puts the user's health or safety at risk because of substance use. Next is an increase in tolerance for a substance. This may seem simple and something that many people without substance use disorders experience, but for those who do have a disorder; this tolerance is more marked and often continues to go up throughout time spent using the substance. Finally, some hallmark criteria for substance use disorders are craving and withdrawal symptoms that occur when a user goes without the substance for a period of time. Craving and withdrawal symptoms are often physically manifested and those with a disorder find themselves soothing these symptoms by using the substance they crave (American Psychiatric Association, 2013). This makes recovery or abstinence from a substance extremely difficult. The DSM-V's new description of substance use disorders increases the clarity of what this diagnosis truly entails.

Although people of any age can receive a diagnosis of a substance use disorder, this paper will focus on diagnosing substance use disorders in adolescents. Lisnov (1999) argues that, "Experimentation with alcohol and other drugs is no longer a characteristic of only a small proportion of youth; rather, it has become the norm among the current generation of American adolescents," (p. 301). The argument can be made that substance use in adolescents is merely a part of development and not necessarily a disorder that requires a diagnosis. Erikson's life stages postulate that during adolescence, youth are in the process of defining their role and identity (McLeod, 2013). Experimenting with substances is a way in which adolescents seek to define themselves. Letcher and Slesnick (2013) add that, "research indicates that substance use, especially alcohol use, is considered a normative activity among college students," (p. 1460).

There is a feeling that the use of alcohol and other substances is simply a part of the social life of adolescents that are often used to cope with stress or relate to an adolescent's peers (Letcher & Slesnick, 2013). The tendency for adolescents to go through this stage of learning about themselves gives way to the trend of using substances. Many of these adolescents experience the criteria of a substance use disorder in their experimentation. However, the question of whether substance use in adolescence is part of development or a disorder remains as both this disorder and life stage are considered.

In considering the diagnosis of substance use disorder in adolescents, attachment theory offers a valuable perspective. John Bowlby played a major role in the development of attachment theory and defined it with several major points. The first of these is that human beings have an innate need to form an attachment to other figures. This is the child's primary bond, often found in mothers. It is added that children need to receive care from their attachment figure during infancy in order to develop successfully. If a child does not receive supportive and effective attachment from their primary figure they may develop difficulties including in disruptive behaviors or in the child's psychology. Finally, attachment theory believes that the caregiver's actions toward the child aid the child in developing his or her sense of self. Therefore, if a caregiver provides positive attachment the child will feel more confident. Conversely if a caregiver reacts to a child primarily with anger the child may also become angry and resistant to forming other attachments (McLeod, 2007). Ball and Legow (1996) add that, "Attachment theory asserts that there is a vital initial stage early in life that lays the foundation for the development of healthy relationships," (p. 535). A positive relationship with a child's primary caregiver gives that child the skills and confidence to create positive and effective relationships later in life.

Attachment theory plays a strong role not only in a young child's sense of self but also in the development of adolescent's identity. One of the biggest parts of the development of an adolescent's sense of self is his or her self-esteem. It is during this life stage that many people struggle to view themselves positively. Securely attached individuals are more successful as they move into their world independently. They tend to be more effective and confident in both interpersonal and intrapersonal relationships (Ball & Legow, 1996). Additionally, a child's perceptions of his orher parent's approval or disapproval are highly correlated with self-esteem (Parker & Benson, 2004). Therefore, if adolescents feel that their primary caregivers, often their parents, view them negatively they are likely to feel negatively about themselves. Self-esteem can make the difference between transitioning successfully through adolescence and facing role confusion.

In regards to substance use, it has been found that self-esteem of an adolescent is closely linked to this behavior. Those with higher self-esteem are less likely to engage in substance use (Letcher & Slesnick, 2013). Cosden and Cortez-Ison (1999) defined a clear relationship between parental attachments, self-esteem, and substance abuse. They found that having negative relationships with caregivers gives way to negative self-esteem, and having negative self-esteem leads to higher rates of substance abuse. Conversely, supportive parenting is linked to higher self-esteem (Parker & Benson, 2004). Additionally, in Parker and Benson's (2004) study, "It was hypothesized that support and monitoring would be associated with higher self-esteem and less risky behavior during adolescence," (p. 519). One of the main criteria for the DSM-V's diagnosis of substance use disorder is engaging in risky behavior. If an adolescent has higher self-esteem and is less likely to engage in risky behavior, he or she may experiment with substances without taking it to the level of falling under the diagnostic category. Self-esteem is associated with many of the aspects of both attachment theory and substance use disorder.

Additionally, other relationships are often formed that mirror an adolescent's relationship with their primary caregivers. Harris, Brazeau, Clarkson, Brownlee, and Rawana (2012) said that adolescents with supportive relationships with their primary caregivers are, "more likely to associate with a positive peer group and therefore be less tempted to use substances," (p. 393). The opposite can also be argued as true. If adolescents have a negative relationship with their primary attachment figure they may be more likely to find themselves in negative peer groups and be *more* tempted to use various substances and perhaps develop a substance use disorder. The occurrence of substance use disorders in adolescents can be linked with their development as defined by attachment theory.

Considering the way in which adolescents develop based on their attachments leads to the closer examination of diagnosing an adolescent with a substance use disorder. Fletcher, Nutton, and Brend (2015) believe that, "traumatic early-childhood experiences and insecure attachments are both independent and interrelated risk factors for developing substance use disorders," (p. 109). The tendency for adolescents who have poor or negative relationships with their primary caregivers to develop substance use disorders points to a deeper issue. Perhaps substance use is not a disorder in itself but rather a symptom of insecurely attached adolescents. Letcher and Slesnick (2013) state that, "individuals with an anxious attachment style may be more likely to use drugs and alcohol," (p. 1460). Based on this statement, the question of whether substance use is a disorder or merely a behavior becomes more obvious.

As previously stated, the use of substances by adolescents is a relatively normal part of development. According to a study by the National Institute on Drug Use (2014a), in the past month, 35.1% of twelfth graders self-reported using marijuana and 37.4% reported using alcohol. With over a third of high school seniors engaging in the use of these substances as well as others, the popularity of experimentation is undeniable. Stated simply, many adolescents engage in substance use without having a diagnosable substance use disorder.

The difference between the youth who experiment with and the youth abuse substances is valuable.

The National Institute on Drug Use (2014b) shows that 42.9% of twelve to seventeen year olds who received treatment for substance abuse were treated for alcohol use and that over 65% received treatment for marijuana use; most of these treatments were received through the juvenile justice system. These high percentages are of those adolescents who pass from experimentation with substances to substance abuse. Experimentation is a part of adolescent development of self-identity and independence but addiction and abuse are extremes that suggest issues within the adolescent that need to be resolved.

With so many adolescents taking substance use to extremes, some form of prevention seems necessary. The simple solution to this would be to implement prevention programs for adolescents with disruptive attachment styles. However, studies have found that prevention programs are often ineffective. In the 1980's and 1990's there was a large increase in school-based drug and alcohol prevention programs. Overall, these programs were found to have, "negative program effect for use of alcohol and cigarettes and no effect for marijuana use," (p.1). This phenomenon that prevention is not effective again points to the question of whether or not substance abuse is truly a disorder or if the use of substances is merely a coping skill for other stressors or diagnosable mental disorders.

If substances are not the primary issue then treatment or prevention of substance use is clearly not effective. Instead, attachment issues should be the focus of treating and preventing substance use disorders in adolescents. Attachment contributes to the development of any individual and is especially important for adolescents. Parents are a particularly important part of development according to attachment theory (Parker & Benson, 2004). School prevention programs for substance use are ineffective because they do not get at the root of the issue.

It is largely believed that attachments can not only be the cause of but also the solution to substance abuse tendencies in adolescents (Ball & Legow, 1996). Attachment models in groups such as Alcohol's Anonymous (AA) have been proven particularly effective for those with substance use disorders (Fletcher, Nutton, & Brend, 2015). The effectiveness of group therapy and support shows that rebuilding positive and productive attachments can counter an individual's reliance on substances. As stated in the DSM-V, the continued use of substances releases chemicals into an individual's brain that give him or her a feeling of pleasure and fulfillment. Forming positive relationships can do the same thing. For example, many substances release dopamine into the brain. Additionally, social situations such as being with peers, playing sports, or receiving physical affection can release the same chemical and create the same feeling of pleasure (National Institute on Drug Use, 2014b). Positive and supportive attachments can provide the same chemical reactions as substance abuse but in a healthy and effective manner.

In addition to receiving these renewed positive attachments from groups, adolescents can grow through participation in individual therapy. Ball and Legow (1996) suggested that an individual can change the way they see their relationships by forming an attachment with a therapist, and that this is an effective form of treatment for substance abuse. In cases of substance addicted adolescents, transference can be used as a part of beneficial treatment. An adolescent may have had difficulties relating to or getting along with adults in his or her life in the past and can work through some of these issues by confronting them with a therapist. If the therapist responds positively and remains supportive, the adolescent can change his or her attachment style.

The final step of treating adolescents with substance abuse tendencies is to rebuild their attachments with their primary caregivers. As previously stated, adolescents who have poor attachments are more likely to use and abuse substances. Repairing these attachments can mean eliminating the initial need or desire to use substances. Although this would be a difficult and long process, rebuilding attachments can change an adolescent's self-esteem, ability to have effective interpersonal and intrapersonal relationships, and their use of substances.

Along with the use of repairing attachments, treatment of adolescents with substance abuse includes a focus on both their strengths and the strengths of their primary attachment figures. Taking a strengths-based perspective in treatment has been proven to be effective (Harris, et. al, 2012). Focusing on not only the individual's strengths but those of their systems and relationships will help rebuild positive attachments and self-esteem. For example if an adolescent has a dysfunctional family and turns to substances it is valuable to look at the good parts of the family dynamic. Perhaps family members share a sense of humor, or a love for cooking, or a dedication to wanting to change. Taking these small moments and making them the focus of treatment allows all family members to feel more secure and supported and therefore gives the adolescent a more positive attachment style.

As attachments are repaired and an adolescent feels more supported, substance use can become the focus. When an adolescent is presenting what appears to be a substance use disorder, it is important to initially put the focus on rebuilding attachments. However, when an adolescent has made positive strides in the area of attachments, his or her use of substances becomes more important. If an adolescent is still abusing substances, it is time to diagnose a substance use disorder. When the four symptoms of a substance use disorder as defined by the DSM-V are still present once other areas of an adolescent's life have been repaired their use of substances changes from a behavior to a disorder. At this point, it becomes clear that the use of substances was never just a behavior that was caused by attachment difficulties, and it may seem that treating attachments did not help with the individual's disorder. However, having supportive and positive attachments during treatment of any mental disorder is immensely helpful. Essentially, substance use disorders may exist in adolescents, but before diagnosing them, attachment issues should be addressed to either act as an alternative form of substance use treatment or simply to build support for treatment of a substance use disorder.

Mental health professionals should not be too quick to diagnose adolescents with substance use disorders because attachment theory shows that there is often a deeper issue that leads to adolescent substance abuse. As adolescents deal with their primary presenting problem of poor attachments, whether or not they still rely on substances should be more closely examined. Once attachment issues are addressed, if reliance on substances still exists, a diagnosis for substance related disorders should be considered, and the positive effects of building or rebuilding attachments can aid an adolescent as they deal with a diagnosed substance use disorder.

References

American Psychiatric Association. (2013). *Diagnostic and statistical manual of mental disorders DSM-5*. Washington, D.C: American Psychiatric Association.

Ball, S. A., & Legow, N. E. (January 01, 1996). Attachment Theory as a Working Model for the Therapist Transitioning from Early to Later Recovery Substance Abuse Treatment. *The American Journal of Drug and Alcohol Abuse, 22,* 4, 533-547.

Cosden, M., & Cortez-Ison, E. (January 01, 1999). Sexual abuse, parental bonding, social support, and program retention for women in substance abuse treatment. *Journal of Substance Abuse Treatment, 16,* 2, 149-55.

Fletcher, K., Nutton, J., & Brend, D. (March 01, 2015). Attachment, A Matter of Substance: The Potential of Attachment Theory in the Treatment of Addictions. *Clinical Social Work Journal, 43,* 1, 109-117.

Harris, N., Brazeau, J. N., Clarkson, A., Brownlee, K., & Rawana, E. P. (January 01, 2012). Adolescents' experiences of a strengths-based treatment program for substance abuse. Journal of Psychoactive Drugs, 44, 5.)

Letcher, A., & Slesnick, N. (July 01, 2013). Romantic attachment, sexual activity, and substance use: findings from substance-using runaway adolescents. *Journal of Applied Social Psychology, 43,* 7, 1459-1467.

Lisnov, L. (February 01, 1999). Adolescents' perceptions of substance abuse prevention strategies. *Sage Family Studies Abstracts, 33,* 130, 301.

McLeod, S. (2007). Bowlby's Attachment Theory. Retrieved April 1, 2015, from http://www.simplypsychology.org/bowlby.html.

McLeod, S. (2013). Erik Erikson. Retrieved April 1, 2015, from http://www.simplypsychology.org/Erik-Erikson.html.

National Institute on Drug Abuse: The Science of Drug Abuse & Addiction. (2014a). DrugFacts: High School and Youth Trends. Retrieved February 18, 2015, from http://www.drugabuse.gov/publications/drugfacts/high-school-youth-trends

National Institute on Drug Abuse: The Science of Drug Abuse & Addiction. (2014b). Principles of Adolescent Substance Use Disorder Treatment: A Research-Based Guide. Retrieved April 15, 2015, from http://www.drugabuse.gov/publications/principles-adolescent-substance-use-disorder-treatment-research-based-guide/introduction.

Parker, J. S., & Benson, M. J. (January 01, 2004). Parent-adolescent relations and adolescent functioning: self-esteem, substance abuse, and delinquency. *Adolescence,39,* 155, 519- 30.

Sloboda, Z., Stephens, R. C., Stephens, P. C., Grey, S. F., Teasdale, B., Hawthorne, R. D., Williams, J., Marquette, J. F. (January 01, 2009). The Adolescent Substance

Abuse Prevention Study: A randomized field trial of a universal substance abuse prevention program. Drug and Alcohol Dependence, 102, 1-3.

The Risk of Anorexia: An Insecure Attachment Perspective

Molly K Watson

Anorexia nervosa is one of the more commonly known eating disorders and can be found listed in the American Psychiatric Association's *Diagnostic and Statistical Manual of Mental Disorders* (5th ed; DSM-5). The most recent version of the DSM was published by the American Psychiatric Association [APA] in 2013 and provides a detailed description of anorexia nervosa, including such information as the diagnostic features, development and course, and diagnostic markers (APA, 2013). The risk factors listed in the DSM-5 include temperamental, environmental, as well as genetic and physiological factors. The DSM-5 fails, however, to discuss the idea that an early, insecure attachment is a very large determiner and risk factor of an anorexia nervosa diagnosis. This paper will discuss the diagnosis of anorexia in terms of attachment theory, making a case for a change to be made in the next published edition of the *Diagnostic and Statistical Manual of Mental Disorders*.

Anorexia Nervosa

Individuals presenting with the DSM-5's diagnostic criteria for anorexia nervosa show three major signs/symptoms. The first is, "Restriction of energy intake relative to requirements, leading to a significantly low body weight in the context of age, sex, developmental trajectory, and physical health" (APA, p. 338). When discussing the idea of low body weight, it is important to understand that the typical anorexia nervosa patient is extremely thin. The DSM-5 outlines the severity of the disorder in terms of a Body Mass Index [BMI]. According to the Center for Disease Control (2015a), a "normal" BMI for an adult is 18.5 – 24.9. The DSM-5 uses BMI ranges to explain the severity level of the disorder. For example, a BMI of equal to or greater than 17 would be considered "mild", while a BMI of less than 15 is considered "extreme" (APA, p. 339). Children and adolescents need to have corresponding percentiles to measure their BMI, which factor in their age and sex (CDC, 2015b).

The second criterion is the extreme fear of putting on weight or getting fat, which also includes behaviors that would support the fear, such as using laxatives or diuretics to keep the weight down (APA, 2013). Weight loss would not be a contributing factor in calming the fear, which enables the client to continue to lose weight without meeting any end goal (APA, 2013).

The third criterion is a distorted perception in the way the client feels about their size and shape (APA, 2013). The DSM-5 also states that this criterion includes a disbelief by the client of the seriousness of their condition (2013). Most people who suffer from anorexia do not see themselves as having any kind of mental disorder; therefore, it is friends and family who often push the individual to seek help (Gabbard, 2014). This can make the disorder incredibly dangerous. In fact, about twelve out of every one hundred thousand individuals diagnosed with anorexia nervosa each year take their own life, one of the highest rates of any disorder in the DSM-5 (Gabbard, 2014; APA, 2013).

While many people with anorexia are able to continue functioning quite well in society, others do not have the capacity to carry out their daily routines in a normal fashion (APA, 2013). The starvation, the lack of nutrients, can cause a great deal of confusion between the mind and the body making it difficult to understand sensations and feelings (Bruch, 2001). It may be just a matter of time before the body's systems become more and more disorganized which may lead to a lower ability to actually function. Bruch explains the perception of some of her patients, saying, "By controlling their eating, some feel for the first time that there is a core to their personality and that they are in touch with their feelings" (2001, p. 4). This seems to contradict what is actually happening in the mind and the body. Their minds appear to be altered by the malnutrition, which may be giving them a false sense of self and a strong sense of empowerment.

Hilde Bruch, a leading psychiatrist in the field of eating disorders, explains her accounts with young women suffering from anorexia. She states that many of the girls confessed to a feeling of inadequacy, saying they felt as if nobody loved them or cared about them (Bruch, 2001). As a way to cope, these girls gained a sense of power from the control they had over their bodies and their weight (Bruch, 2001). This sense of power gave them the idea that they mattered; that they were important. One of Bruch's patients said, "By losing weight, accumulating empty pounds, I would give myself permission to be nurtured, to be cared for, to be recognized" (Bruch, p. 5). Another of her patients claimed that she would lose her mother's love if she ate (Bruch, 2001).

What leads a young girl to turn to starvation? The DSM-5 discusses only a few risk factors that may be involved. The authors go into detail explaining the various "genetic and physiological" risk factors, such as genetic implications to the disorder and brain abnormalities (APA, 2013). They also state environmental factors such as affiliation with cultures, settings and occupations, which value small frames and the idea of being thin (APA, 2013). The DSM-5 even offers up temperamental risk and prognostic factors, such as anxiety disorders and obsessional traits, as factors that could later contribute to an onset of anorexia nervosa (APA, 2013). What the authors failed to incorporate, which seems to be obvious to many researchers and clinicians in the field, is the factor of early, unstable attachments.

Attachment

Attachment theory has been influencing clinicians and experts in the field of psychology and human development for years. John Bowlby and Mary Ainsworth worked to help define attachment and its impact on infants and adults, which helped to develop the categories of secure and insecure attachments that are used today (Tasca & Balfour, 2014). The secure category of attachment is associated with a healthy bond between child and mother, one that encourages independence and the ability to form and maintain close relationships with others (Eggert, Levendosky, Klump, 2007). Infants who have caregivers who are emotionally available develop a healthy autonomy and see themselves as important and loved (Troisi, et al., 2006).

Researchers have found that mothers who cared for their infants according to their needs more than the actual needs of the baby were more likely to establish an unhealthy autonomy and self-concept (Gabbard, 2014). These are considered to be insecure attachments. Infants who do not have emotionally available caregivers tend to have lower self-esteem, more psychopathology, and lower academic achievement (Eggert, Levendosky, Klump, 2007). Potential psychopathology includes disorders such as anorexia nervosa.

Many researchers have developed different attachment models, however, they all seem to follow a general foundation of belief that if the child's emotional needs are not met, then he/she will have difficulty feeling secure and safe later on in life. In order to fill this emotional gap, they may find themselves behaving in ways that are unhealthy and/or destructive. Hooper and Dallos (2012) explain the importance of the Crittenden's Dynamic Maturational Model (DMM), which "locates behavioral and psychiatric disorders into the context of family-attachment relationships" (p. 454). The model explains that a child is more likely to go to extremes to find comfort in an environment in which they feel very unsafe and threatened (Hooper & Dallos, 2012). These efforts become coping mechanisms of defense for the individual and can prove to be harmful later on in life. Anorexia nervosa could be an example of these coping mechanisms.

Anorexia and Attachment

The preoccupation with weight and food may resemble a much larger issue that the anorexic client is dealing with. Researchers have found there to be a disruption with the anorexic client's sense of self (Gabbard, 2014). Gabbard (2014) explains:

> Most patients with anorexia nervosa have a strong conviction that they are utterly powerless and ineffective. The illness often occurs in 'good girls' who have spent their lives trying to please their parents only to suddenly become stubborn and negativistic in adolescence. The body is often experienced as separate from the self, as though it belongs to the parents. These patients lack any sense of autonomy to the point that they do not feel in control of their bodily functions. (p. 359)

Gabbard (2014) also explains that this lack of self-concept and autonomy can often be traced back to early, unstable parental attachments. He claims that children who are not validated through a caregiver's response will often lack a healthy autonomy, feeling like they are not the center of their own being (2014). An anorexic patient may just be trying to validate his or her own self-worth and admiration from his/her caregivers (Gabbard, 2014). Hooper and Dallos (2012) state that it is most important for a child to feel validated during the years when he is forming his identity. "Specifically, in relation to eating disorders, the family plays a crucial role in the development of the child's identity and beliefs regarding their own body" (Hooper & Dallos, p. 453).

Many researchers have looked at the connection between early childhood attachments and the onset of anorexia nervosa and/or bulimia nervosa later on in life. In a study of 65 women diagnosed with bulimia and 31 women diagnosed with anorexia, the women were given questionnaires to assess their body satisfaction, their level of separation anxiety, level of depression, as well as their respective attachment styles (Troisi et al., 2006). The study showed connections between body dissatisfaction, insecure attachment and early separation anxiety. Troisi et al. (2006) state, "body dissatisfaction was strongly associated with early separation anxiety and an insecure style of attachment" (p. 452). They go on to conclude, "There is mounting evidence that elevated body dissatisfaction is a risk factor for eating disorders and that this relation is mediated by increases in dieting and negative affect" (Troise et al., p. 453).

Tasca and Balfour looked at some of the current research on eating disorders and attachment and analyzed a variety of studies, keeping their eyes on clinical studies of subjects with eating disorders. To simplify the process of reviewing the current research, Tasca and Balfour (2014) looked at major themes in regards to attachment and eating disorders; some of which include, "the prevalence of attachment insecurity and the level of reflective functioning, the association between attachment insecurity and eating disorder diagnosis or symptom severity, and associations with trauma and disorganized mental states" (p. 712). Tasca and Balfour's findings shed light on the role attachment plays in anorexia.

In Tasca and Balfour's first theme of "attachment insecurity and reflective functioning," they found insecure attachments to be prevalent in 70-100% of the samples with subjects who had eating disorders (2014). They also found that subjects with eating disorders had a lower capacity to make sense of themselves and others, and suggested that this may be a "specific feature" of the diagnosis (Tasca & Balfour, 2014).

When looking at "attachment insecurity and eating disorder diagnosis or severity," they found inconsistent findings with the specific diagnosis and the category of attachment, however, they do conclude that body dissatisfaction is very much associated with a "need for approval," which is connected to attachment (Tasca & Balfour, 2014). Tasca and Balfour (2014) state, "Preoccupation with relationships and fear of abandonment, especially when expressed as needing others' approval, may be a particularly problematic attachment-related insecurity that may put individuals at risk for greater disorder symptom severity" (p. 713).

One of the more important findings of Tasca and Balfour's meta-analysis is their direct association between trauma, loss and abuse during childhood and the psychopathology of eating disorders (2014). Insecure attachments could make it extremely difficult to cope with the stress of a traumatic event like abuse or loss. Although trauma is difficult for any child, those with secure attachments have a better chance of regulating their emotions so they can effectively respond to stressors (Tasca & Balfour, 2014). Hooper and Dallos (2012) quite poignantly explain that many children who go through some kind of trauma or abuse may fear their parents while still needing them a great deal.

Father-Daughter Relationship

When discussing attachment, the mother-child relationship tends to get the most attention versus any other adult-child relationship. As mentioned earlier, the DSM-5 states that females are at a higher risk of developing an eating disorder, and the onset typically occurs during adolescence (APA, 2013). If the research is pointing to insecure attachments as a risk factor for these females, it is necessary to investigate all attachment relationships they may have, including the relationship they have established with their fathers.

When a young girl hits puberty, her body starts to change in drastic ways. She may feel as though she has lost control of something that she once saw as a cooperative part of her being. During this time, I think it is safe to say that mothers are typically involved more, as they have experienced these changes themselves and may have a better idea of what their daughters need during this time in their lives. Where are the fathers during this time? What happens if a father does not establish an emotionally supportive bond with his daughter during infancy and that insecure attachment perpetuates into preadolescence? Once that child hits adolescence, would it also be safe to argue that those same fathers may emotionally pull back even further? As her body changes and she continues to grow into a woman, it is important for a girl to feel acknowledged and validated by possibly the most important man in her life, her father (Hooper & Dallos, 2012). Investigating the father-daughter relationship further and its possible impact on the onset of anorexia is extremely important.

Only a few researchers have set out to find more information about this father-daughter relationship. Hooper and Dallos (2012) used three men and their daughters to learn more about attachment and the role that fathers play in their daughter's development of the disorder in their study.

Hooper and Dallos used the Adult Attachment Interview as well as unstructured interviews with both the fathers and the daughters (2012). One consistent finding in both the men and the women was their difficulty to talk about their feelings and their relationships, especially when it came to painful topics (Hooper & Dallos, 2012). Hooper and Dallos (2012) suggest that the women may actually be communicating their feelings through the action of self-starvation. The authors also hypothesized that the fathers may not have been in a position to help their daughters during stressful times, which led the daughters to experience feelings of rejection (Hooper & Dallos, 2012). The rejection could cause the girls to turn towards their mothers, or an alternate source of comfort, if another stressful or traumatic event should occur (Hooper & Dallos, 2012). "In order to increase parental behavior, children may unconsciously learn that they cannot display their needs in full, rounded and unabridged form. Instead they develop secondary behavior strategies in an attempt to recover parental interest and availability (Hooper & Dallos, p. 464)." They also conclude, "If emotional distress is not able to be voiced or heard, the girls may require this pain to come out in a different form, such as food control" (Hooper & Dallos, p. 464).

While this study was extremely small, their findings are consistent with other research in pointing out the connection between attachment and eating disorders. Hooper and Dallos' work sheds light on the father-daughter attachment, which has been a relatively new focus for researchers. Their findings, while hard to generalize, show a direct correlation between an insecure father-daughter attachment and the onset of anorexia. More research is needed in this area, with larger, more generalizable samples to further explore this idea.

Conclusion

Extensive research has shown there to be a significant relationship between insecure attachment and the onset of anorexia nervosa. During a time when identity and autonomy is being explored, girls are in need of that comfort and emotional support to help them navigate through the transition into young adulthood. Without this support, they may be left to find security in the form of self-starvation.

The risk factor of an insecure attachment is not listed in the DSM-5 under the anorexia nervosa diagnosis code and information. If the authors of the DSM-5 did not feel that there was enough research to support this risk factor, I feel confident that they will change their minds before the DSM-6 is ready to be published. It seems as though new research on the topic is being conducted every year, proving the connection between attachment and anorexia. While research in abundant in this area, additional research is needed in the area of anorexia and attachments with fathers, specifically. These findings could greatly assist therapists as they try to understand what it is like for girls living with anorexia, and might possibly lead them to appropriate treatment modalities.

References

American Psychiatric Association. (2013). *Diagnostic And Statistical Manual Of Mental Disorders* (5th ed.). Washington, DC: pp. 338-345.

Bruch, Hilde. (2001). *The Golden Care: The Enigma Of Anorexia Nervosa.* Cambridge, MA: Harvard University Press.

Center for Disease Control. (2015a). About BMI for adults. Retrieved from:http://www.cdc.gov/healthyweight/assessing/bmi/adult_bmi/index.html

Center for Disease Control. (2015b). About BMI for children and teens. Retrieved from: http://www.cdc.gov/healthyweight/assessing/bmi/childrens_bmi/about_childrens_bmi.html

Eggert, J., Levendosky, A., Klump, K. (2007). Relationships among attachment styles, personality characteristics, and disordered eating. *International Journal of Eating Disorders. 40*(2), pp. 149-155.

Gabbard, G. (2014). Substance-related and addictive disorders and eating disorders. *In Psychodynamic Psychiatry in Clinical Practice (5th ed.).* Washington DC: American Psychiatric Publishing. pp. 357-366

Hooper, A., and Dallos, R. (2012). Fathers and daughters: Their relationship and attachment themes in the shadow of an eating disorder. Contemporary Family Therapy: *An International Journal. 34*(4), pp. 452-467.

Tasca, G.A., and Balfour, L. (2014). Attachment and eating disorders: A review of current research. *International Journal of Eating Disorders*, 47(7), pp. 710-717

Troisi, A., Di Lorenzo, G., Alcini, S., Nanni, Roberta C., Di Pasquale, C., Siracusano, A. (2006). Body dissatisfaction in women with eating disorders: Relationship to early separation anxiety and insecure attachment. *Psychosomatic Medicine*, 68(3), pp. 449-45

Major Depressive Disorder and the DSM V:
A Diagnostic Critique

Shatara Johnson

History of DSM

The DSM 5 (Diagnostic and Statistical Manuel of Mental Disorders) is a highly referenced manual utilized by many mental health professionals, educators, and students. In the early 1840's, the DSM was initially created as a census to collect statistical information on mental disorders and their frequency within American society. By the late 1800's the statistical census revealed seven categories of mental health disorders representing idiocy and insanity which included: mania, melancholia, monomania, paresis, dementia, dipsomania, and epilepsy (APA, http://www.psychiatry.org/practice/dsm/dsm-history-of-the-manual). Currently the DSM 5 contains over one hundred mental disorders including their qualifying criteria which have all been influenced by the APA and many other committees. Although the DSM 5 deems to be very helpful as it adheres to diagnosing individuals with mental disorders comparable with societal norms, it fails to offer a solution that tends to the holistic person. The DSM 5 emphasizes both biological and psychological determinants that can lead to each mental disorder, but only provides a solution that focuses on the biological determinants. To consider one's biological contributors and not the environmental stimuli affecting an individual's psychological well-being preferences one contributing factor over the other. Highlighting an individual's internal mechanisms as being flawed or atypical forces them to believe that they are the problem and need to be fixed. On the other hand, considering the impact of the environment on mental disorders would not only lead to a significantly lower rate of misdiagnosis, but will also impel clinicians to tackle every aspect of the person before primarily considering harmful psychotropics.

DSM & Major Depressive Disorder

Major Depressive Disorder deems to be an example of a mental disorder demonstrating strong biological and psychological contributors to the diagnosis. The criteria for Major Depressive disorder are characterized by discrete episodes lasting at least two weeks or longer (DSM 5, 2014, p155). In order to be diagnosed with Major Depressive Disorder an individual must display at least five symptoms including: depressed mood most of the day, diminished interest of pleasure, significant weight loss, insomnia or hypersomnia, psycho-motor agitation, fatigue or loss of energy, feelings of worthlessness, diminished ability to concentrate and recurrent thoughts of death (DSM 5, 2014, p161). All of these symptoms are pertinent to the diagnosis, but the DSM 5 states that it is required for one of the five symptoms to include either a depressed mood or loss of interest in pleasure (DSM 5, 2014, p 160). When considering the biological aspects of the mental disorder, Gabbard's referenced studies demonstrate little to no substantial evidence supporting biological contributors of the disorder. According to Kendler and his colleagues (1993) they followed 680 female twin pairs to determine if an etiological model could be composed to predict major depressive episodes. It was concluded that the role of genetic factors was substantial but not tremendous. Gabbard states (2014) that trauma in early development can lead to permanent biological transformations. For example, a study done by Vythilingam et al. (2002) found that depressed women with an experience of early child hood trauma had 18% smaller mean left hippocampal volume than healthy subjects. Last but not least, a study noted by Gabbard documents higher levels of CRF (corticotrophin releasing factor) in patients with depression as opposed to healthy patients (DSM 5, 2014, p 220). Gabbard reflects (2014) "We now understand the etiology of unipolar depression as approximately 40% genetic and 60% environmental" (p219). It is quite evident that there is not a significant amount of evidence to strongly support that Major Depressive Disorder is predominantly a biological disorder.

Major Depressive Symptoms & Stress

Redirecting the focus of Major Depressive disorder from biological to probable environmental contributors of the disorder broadens the scope of the criteria and considers how the environment can negatively affect an individual's mood. In a literature review done by Anakwenze and Zuberi (2013) urbanization is a major contributing factor to symptoms associated with mental disorders (p 147). The authors suggest that individuals living in the city have a higher risk for mood disorders. Anakwenze and Zuberi (2013) go on to estimate that at least 46 percent of the individuals living in the city will experience at least one disorder in the DSM 5. It is important to look at individuals residing in the city because there are more marginalized populations dealing with life stressors influenced by spatial segregation along racial, ethnic, or socioeconomic lines that reinforce poverty for low-income residents (Anakwenze & Zuberi, 2013, p 148). Segregation acts as a hindrance to low-income populations because it does not allow the formation of social networks that would result in employment or social mobility. Anakwenze and Zuberi assert (2013), "Unstable work and low income decrease one's perceived self-efficacy" (p 148). Economic pressures caused by unstable work and low income can create feelings of emotional distress and as a result tend to lower an individual's sense of efficacy as it relates to having positive influence over their families and the environment (Anakwenze & Zuberi, 2013, p 149). Living in a neighborhood that offers a lack of social support, high crime and insecurity and enhanced victimization can further create symptoms of depression because of the exposure to an excessive amount of traumatic events. This article demonstrates that low-income populations living in the urban city are more prone to experiencing unfortunate events that can lead to symptoms of depression, due to a lack of financial stability, social support, and social mobility. Not being able to positively impact one's own future or others that are close to you only forces the individual to internalize their feelings of adversity.

A study researched by Dunn et al. further examines the relationship between schools and neighborhoods as risk factors for depressive symptoms. Dunn and his colleagues proclaim that it is important to consider the role of the schools because 95% of the nation's young people for approximately 6 hours per day and at least 11 continuous years of their lives are enrolled in school (Dunn et al, 2015, p 732). Schools are also well-defined social institutions that provide access to a variety of supportive relationships influencing mental health. On the contrary neighborhoods promote unstructured social activity outside of school hours and during vacation time for the students. The neighborhood also influences a parent's capacity to raise their children through shaping community norms, supervision and monitoring, collective efficacy, and reductions in the burdens and stressors associated with caregiving (Dunn et al, 2015, p 732). This study analyzed data from the National Longitudinal Study of Adolescent Health with cross classified multi-level modeling to examine between-level variation and individual, schools, and neighborhood level predictors of adolescent depressive symptoms (Dunn et al, 2015, p 733). The results indicated that schools appeared to drive the between-level variance in depressive symptoms more than neighborhoods. Schools may have a unique potential to affect, at a population level, the prevalence of depression among adolescents (Dunn et al, 2015, p 738). Revealing that schools have a great influence on an individual's mental health and development of depressive symptoms illustrates that socialized institutional learning and various social relationships encountered during school hours can have a greater impact on an individual than their own unstructured neighborhood; hence it is fair to say that state imposed regulations such as academia, can also act as a detriment to an individual's mental health.

McTerman, Dollard & LaMontagne inquire about the relationship between the workplace and workers' experience of depressive symptoms. McTerman states (2013) psychosocial risk factors such as job strain and bullying can result in a decrease in work productivity. McTerman et al. (2013) write, "Job strain is a combination of high job demands and low job control, which is the most deleterious situation for a worker's health" (p 323). When workers are faced with high demands this causes a flight or fight response that enables the worker to cope for the moment or in a short term period. This can be problematic because if such high demands become persistent and the employee lacks the appropriate level of job control and social support, the aroused energy experienced by the employee evolves into stress that affects the individual physically and mentally. The stress of the work place will eventually lead to low energy and fatigue. In conjunction with job strain, bullying such as belittling, singling out and isolation of a targeted co-worker can also act as a high risk factor for depressive symptoms. Bullying can directly affect one's own view of self-image or perceived self. If policies for bullying prevention and management are lacking, this may serve to reinforce or even foster bullying (McTerman et al, 2013, p 324). The work place should be a reliable source of income and create incentives for the worker to have the desire to remain productive and cooperative. When an employee does not feel a sense of security, positive feed-back, and overall care, it can result in withdrawal from the job and feelings of conflict and stress.

Choi & Mcdougall took it a bit further and researched depressive symptoms between homebound older adults and ambulatory older adults. It is hypothesized that many older adults are vulnerable to health conditions later in life that can lead to a lack of mobility. This study relied on self-reported data to measure depressive symptoms, health-related stressors, other life stressors, and coping resources (Choi & McDougall, 2007, p 312-313). The results of this study revealed that 42% of the homebound group compared to 13% of the senior center participants, scored 5 or higher on the GDS. In other words, the homebound adults displayed more depressive symptoms than the ambulatory adults. The lack of mobility amongst the homebound older adults can create social isolation imposed by chronic illness and functional limitations, which increases their vulnerability to depressive symptoms. Choi & McDougall acknowledge (2007) their homebound state makes their mental needs unrecognized which is a barrier to their receiving appropriate treatment (p 310). Also homebound older adults require in-home support services for their IADL (instrumental activities of daily life) and ADL (activities of daily living) tasks and if these needs are not properly addressed due to a lack of social services or care from a nursing facility these individuals are likely to have limited effects on their quality of life (Choi & McDougall, 2007, p 310). This is deemed to be vital because if a homebound older adult does not have access to social activities and self-help this could enhance feelings of helplessness and self-pity which could result in depressive symptoms.

Last but not least, Graham et al reflected on factors influencing depression among BSMM (black sexual minority men). Using an observational cross-sectional design and self-administered online surveys, Graham advises (2011) BSMM experience more depressive symptoms and anxiety than their male heterosexual and black female counterparts and at minimum parallel to those of their white sexual minority counterparts (Graham et al, 2011, p1). Exploring the psychosocial health of BSMM; Graham showed that they are challenged with developing an acceptable identity racially and sexually (Graham et al, 2011, p 2). During this time of exploration and seeking acceptance, BSMM experienced discrimination and harassment on a daily basis. Participants in the study expressed experiencing the most discrimination and harassment in public places, retail settings, and the criminal justice system (Graham et al, 2011, p 4). Dealing with discrimination and harassment unquestionably leads an individual to feelings of being ostracized, eccentric and overall unfit for societal norms. This unhealthy thought process can also lead to lower self-esteem and cause relentless distress.

Systems Theory

After reviewing qualitative research on life stressors that can influence depressive symptoms in various contexts, systems theory helps to clarify the dyadic relationship between the individual and his or her environment. The ecological perspective explains adaptation and how it relates to the environment. Adaptation refers to an individual's ability to reproduce and survive in a given habitat. We are able to adapt not only by our predisposed skills or inherited genetic traits, but also by using the assets of the environment we are placed in. There are two different environments: the natural environment and the social environment. The natural environment constitutes the actual geographical location, and the social environment represents the network of relationships of individuals and groups (Robbins et al, 2012, p 32). Within an environment are ecosystems on a Micro, Meso, and Macro level. These ecosystems allow fluid or rigid reciprocal relationships between an individual and his or her environment. Micro level relationships represent the self, family members, and close friends, while the Meso level and macro level include larger entities for instance, community, organization, state, nation, etc (Robbins et al, 2012, p. 33). Throughout this exchange of relationships between ecosystems; dominance or succession may take place. Dominance involves one group of individuals in the same natural environment seizing control or power over the more vulnerable group, leaving the vulnerable group with a lack of or limited amount of resources. Robbins theorizes (2012) that an environment could either support or fail to support the adaptive achievements of autonomy, competence, identity formation, and relatedness to others. When the environment and the individual are not able to reciprocally adapt to one another--this reflects the "Goodness of fit" (Robbins et al 2012, p33). For example, if an individual is striving for resiliency but the resources available in the environment do not allow them to achieve desired success, then this results in a disturbance within the "Goodness of fit," meaning that the environment is no longer beneficial to the individual and his or her desired happiness or success. Robbins states (2012) "When inputs or stimuli are insufficient, excessive, or missing altogether, an upset occurs in the adaptive balance, which is conceptualized as stress: the usual 'fit' between the person and environment has broken down" (p. 34). According to Robbins (2012) stress is a psychosocial state of being generated by discrepancies between needs and capacities and environmental qualities. Stress arises in life transition, environmental pressures, and interpersonal processes. In conclusion, adapting to the environment is necessary for survival but if you are not in an environment that allows for social mobility, then the harder it will be to reach success, which can create depressive symptoms associated with Major Depressive disorder.

Systems Theory and Major Depressive Disorder

Major Depressive disorder specifies the diagnostic criteria of having a depressed mood most of the day, diminished interest of pleasure, significant weight loss, insomnia or hypersomnia, psychomotor agitation, fatigue or loss of energy, feelings of worthlessness, diminished ability to concentrate and last but not least recurrent thoughts of death (DSM, 2014, p161). All of the criteria listed above exemplify depressive symptoms as they relates to factors of the environment. It is quite evident that life stressors caused by environmental factors can lead to such internal feelings of withdrawal, sadness, discouragement and many more emotions representing dissatisfaction with one's current situation. Sharpley and Bitsika (2010) proclaim that symptoms of depression can be seen as a coping mechanism to deal with life stressors. Life stressors cause unwarranted discomfort for the individual learning to cope with adversity; therefore, going into solitary confinement or distancing one's self from all perceived stressors allows the individual to feel as if they are temporarily taking themselves away from the problem (Sharpley & Bitsika, 2010, p4). It may not be positively adapting to one's situation but with a lack of social support and guidance, this may appear to be the best solution for individuals experiencing such adversity.

Further Research

In conclusion, the number of patients being diagnosed with Major Depressive Disorder as defined in the DSM 5 criteria increases by 20% each year (Health line, 2012, http://www.healthline.com/health /depression/statistics-infographic). This is problematic because each individual being diagnosed with

Major Depressive Disorder is immediately prescribed psychotropic drugs. Prescribing antidepressants does not permanently fix the problem. Antidepressants have not been proven to be 100% effective, and they do not address the problems of the environment that directly affect the person. They enhance the mood of the individual and create a dependence on the medication (Yen et al, 2009, p1038). Gabbard states that 60% of the contributing factors of major Depressive Disorder are environmental. If that is a valid fact, then why is medication a primary solution instead of making changes to the environment? Medications are only geared toward the internal person but the external contributors that impact the person's ecosystems are not being addressed or ignored. Instead of labeling an individual as having Major Depressive disorder why not primarily consider making improvements to marginalized communities, resources, worker benefits, social support groups, and various iconic symbols publicized on social media? It would be advantageous to the individual to address all aspects of his or her life, for instance, biological makeup and the environmental challenges that one may face. Not considering the entire person in the environment and how it impacts the individual could lead to over diagnosing adults as well as children. This will also reiterate to the individual that the problem lies within his or her self, and medication is the only solution. If an individual is constantly labeled as having a mental disorder, it is likely that he or she will perceive him or herself as the problem, which can lead to further depressive symptoms such as hopelessness. It is important to affirm that the individual is not the problem, and emphasis should not be placed on medications as a long-term treatment. In fact society should be held accountable for creating social stratification systems that allow for binaries and categorization of human beings. Being able to diagnose an individual can be beneficial because it allows a person to have clarity if they are troubled by such symptoms and it helps to normalize a specified disorder but, on the contrary, it will be more beneficial if an individual's environment was taken into consideration before immediately resorting to antidepressants. If medications are instantly suggested, then how will one know that the individual's depressive symptoms would not have decreased with an environment that provided a 'goodness of fit'? Perhaps medication was not needed at all? This remains an on-going issue that will never be addressed as long as medication continues to be the first response to individuals experiencing depressive symptoms.

References

Robbins, S.P., Chatterjee, P., Canda, R. E., (2012). Contemporary Human Behavior Theory. *A critical Perspective for social work*, 3, 1-482.

APA. (2013) Diagnostic and statistical mental disorders. DSM, 5, 1-946.

Gabbard, O,. Psychodynamic Psychiatry in clinical practice. *DSM 5 Edition*, 5, 1-639

Sharpley, F,. Bitsika, V,. (2010). Is Depression Evolutionary or Just Adaptive? A comment. *Depression research and treatment*, 1-8.

Graham, F,. Aronson, E,. Nichols, T,. Stephens, F,. Rhodes, S,. (2011) Factors influencing depression and anxiety among black sexual minority. *Depression research and treatment*, 1-9.

Choi, G,. McDougall, J,. (2007). Comparison of depressive symptoms between homebound older adults and ambulatory older adults, Aging & Mental health, 310-322.

McTernan, P,. Dollard, F,. LaMontagne, D,. (2013). Depression in the workplace: An economic cost analysis of depression-related productivity loss attributable to job strain and bullying. *Work and stress*, 321-338.

Dunn, c,. Milliren, E,. Evans, R,. Subramanian, V,. Richomond,. (2015) Disentangling the Relative Influence of Schools and Neighborhoods on Adolescents' Risk for Depressive Symptoms. Research and Practice, 732-740.

Anakwenze, U,. Zubri, D,. (2013). Mental Health and Poverty in the Inner City. *National Association of Social Workers*, 147-157.

Yen, F,. Chen, C,. Lee, Y,. Tang, T,. Ko, H,. Yen, Y,. (2009) Association between quality of life and self-stigma, insight, and adverse effect of medication in patients with depressive disorders. Depression and Anxiety, 1033-1039.

Well, T,. Clerkin, M,. Ellis, J,. Beevers, G,. (2014). Effect of Antidepressant Medication Use on Emotional Information Processing in Major Depression. *Psychiatry*, 195-199.

Cannabis Use Disorder as a Social Construction: Implications for Social Work

Sarah Leight

Cannabis use has appeared in every edition of the DSM, categorized by substance abuse. Yet American society has been aware of marijuana since before the publication of the first DSM in 1952 (http://www.psychiatry.org/practice/dsm/dsm-history-of-the-manual_). But perhaps we have forgotten how old marijuana use actually is, and marijuana has existed for far longer than modern day civilization, dating back to the 4th millennium BCE (Merlin, 2003, p. 312). Marijuana's identity in the U.S. today is not of a plant, but a powerful and controversial drug, widely available yet socially and legally questionable. Cannabis use disorder, as it appears in the DSM V (American Psychiatric Association, 2013) is the result of the powerful and addictive nature of marijuana, which affects its users' cognition, mental state, and physical health. Looking closer however, cannabis use disorder is not a substance abuse disorder, but merely a social construct resulting from decades of cultural and social discrimination, legal inequity, and medical inaccuracy. Using a social constructionist perspective, one can explore the evolution and nature behind the truth of cannabis use disorder by thoroughly examining history and the subjective reality of our existence.

Marijuana as an Ancient Plant

Exploring cannabis use disorder as a social construct must begin with an analysis of the marijuana plant. Cannabis has been grown, cultivated, and processed for the last five thousand years. The plant is from the Cannabaceae family and is native to Asia. The plant was primarily grown worldwide as a fiber crop, as the hemp seed from the marijuana plant can be produced to make fiber for clothing and paper (Stewart, 2009, p. 109-110). The marijuana plant was also used for medicinal purposes. "Egyptian, Chinese (2700 BC) and Assyrian (800 BC) sources indicate that [marijuana] is one of the oldest drugs in history (Mechoulam & Eeigenbaum,1987). The earliest reference to the medicinal properties of cannabis dates back to 2700 BC (Grinspoon & Bakalar,

---

1993)" (Adams & Martin, 1996, p. 1586). Besides its medical and agricultural importance, marijuana has also historically been used for its psychotropic properties—marijuana as an intoxicant is not a modern day invention. "…The euphoric properties of cannabis were discovered in India between 2000 and 1400 BC…" (Adams & Martin, 1996, p. 1586). Cannabis in this context was principally used ritualistically and some historians argue that early cannabis use yielded the inspiration for the first religious concepts (Merlin, 2003, p. 295).

There is no evidence at these origins, that marijuana use as an intoxicant was a moral or social issue, and it was openly used, as is illustrated by the discoveries of ancient artifacts, including smoking tools, and also ancient writings (Merlin, 2003, p. 313). Cannabis in all its forms was used regularly throughout the Mediterranean and Asia, and would then eventually move to the African continent, but not for another 2000 years. Its appearance in Western Europe also occurred later, and marijuana's eventual move into the South and North American continent did not occur until after that. The reports and research on marijuana in the Americas is diverse, but for our purposes we can agree with the research of Merlin and state that "Cannabis did not appear on the North American continent until after 1492," (Merlin, 2003, p. 316) or the beginning of European colonization and expeditions.

Marijuana Law and Regulation in the U.S.

Moving forward through industrialized civilization, marijuana continued to be used for medicinal, recreational, and agricultural purposes, with no repercussions. Up until the 20th century, the U.S. did not regulate the use of marijuana. Then in 1914, El Paso, Texas saw the first "…local ordinance banning the sale or possession of marijuana…" (Warf, 2014, p. 429). Moving into the 1920s, the U.S. experienced a counterculture movement which involved non-white and urban marijuana users.

> The 1920s brought new rounds of users, including African-Americans and the Greenwich Village bohemian community (Polsky 1967). Sailors and Caribbean immigrants brought marijuana to coastal cities…Circuits of jazz musicians carried the drug to St. Louis, Kansas City, Chicago, and Harlem…American jazz,

increasingly popular in Britain, also facilitated the trans-Atlantic diffusion of the drug. (Warf, 2014, p. 429)

At this time, U.S. law did not distinguish between the strain of cannabis that produced hemp for fiber, and the cannabis strain that produced buds with THC. Due to this, the hemp plant was outlawed in most states by 1931 (Warf, 2013, p. 429).

During this era, marijuana law and social interpretations of marijuana use were encouraged by the first commissioner of the Federal Bureau of Narcotics, Harry Anslinger. Anslinger encouraged Congress to then pass the Marijuana Tax Act in 1937, "…which put cannabis under the regulation of the Drug Enforcement Agency, effectively criminalizing possession throughout the country (Canada followed in 1938)" (Warf, 2014, p. 430). The re-legalization of hemp occurred during WWII, with the government careful to distinguish the hemp strain of cannabis from the strain that produced the intoxicating drug. Then came an onslaught of legislation.

> Following the war…antihemp programs initiated by the DEA required permits to grow the plant, and in 1948 it was criminalized again…In 1951, Congress passed the Boggs Act, which specified the same penalties for marijuana possession as for heroin (Schlosser 2003). (Warf, 2014, p. 430)

The Boggs Act "provided uniform penalties" (Sloman, 1979, p. 189) for the possession and sale of marijuana and all other narcotics. Then, in 1972,

> President Nixon appointed the National Commission on Marijuana and Drug Abuse, which soon concluded that [marijuana] should be decriminalized; evidence notwithstanding, [Nixon] immediately rejected their findings. Nixon's departure from office, however, was followed by a steady movement toward legalization…Eleven states essentially decriminalized small amounts of the drug…Legalization was supported by the American Bar Association, the American Medical Association, the National Council of Churches, and President Jimmy Carter (Schlosser 2003). (Warf, 2014, 431)

At this time, the majority of marijuana smoked in the U.S., came to the country from outside of the states; Mexico, Colombia, Canada and Jamaica were the primary suppliers for U.S. buyers (Warf, 2014, p. 431). A few years earlier, the U.S. government "launched Operation Intercept along the border with Mexico, ostensibly to reduce the inflow of drugs" (Warf, 2014, p. 431). This Intercept included the 1975 "large-scale spraying of the herbicide paraquat over Mexican marijuana fields" (Warf, 2014, p. 431). However, "as the drug's supply declined and prices rose, a domestic industry arose in response" and U.S. growers started cultivating the marijuana plant indoors (Warf, 2014, p. 431). Growers learned to produce plants with higher contents of THC which offset the cost of their production (Warf, 2014, p. 431)

By the 1980s, the introduction of the political New Right halted the previous decriminalization movement.

> In 1979, the Drug Enforcement Agency initiated the Cannabis Eradication/Suppression Program, focusing on California and Hawaii. In 1982, President Reagan launched a war on drugs, including the White House Drug Abuse Policy Office...Courts were encouraged to adopt mechanistic sentencing formulas, simplistic "zero tolerance" legislation...(Warf, 2014, p. 432)

Passing through the 1990's and into our 21st century, our federal government "still classifies marijuana as a Schedule I controlled substance" in the same category as heroin, and LSD (Warf, 2014, p. 432). This classification specifies that marijuana has a "high potential for abuse and addiction, no accepted medical uses, and no safe level of use" (Warf, 2014, p. 432). Despite this classification, politically, marijuana law is still changing. As of 2015, 4 states and the District of Columbia have legalized the recreational use of marijuana, 19 states have allowed the sale of medical marijuana, and 14 states have decriminalized marijuana use (http://norml.org/laws).

This outline of marijuana law attempts to show the unbiased and simplified timeline of marijuana history, devoid of the social implications of the creation, implementation and consequences of marijuana law. Attempting to separate the seemingly "objective" nature of marijuana law before using a social constructionist perspective assists in understanding the true nature of marijuana law, which both reflects and affects society's changing opinion of the drug. Although marijuana law is not explicitly linked to the publication of cannabis use disorder, exploring marijuana law illustrates how marijuana use is a social construction, and cannabis use disorder is the result of a trickle down effect of marijuana regulation. In this manner, both marijuana law and cannabis use disorder are social constructs.

Social Constructionism

Social constructionism was born in the 20th century as a response to "mainstream social science" which theorists viewed as the "manifestation of the one-sided individualism and techniques of modern times" (Robbins, 2012, p. 328) Simply put, social constructionism is the concept that our reality is subjective and not objective. Objective reality and the means by which it is organized, is what social constructionists observe as a farce and instead propose that the "unchanging foundations for the human sciences...metaphysical basis or moral standard for judging our beliefs" are "...simply another historically influenced interpretation, a mere projection of our particular community's viewpoint onto the universe. [Social constructionists] contend that it is time to acknowledge the fundamental truth that all our beliefs and values are strictly relative" (Robbins, 2012, p. 329) Social constructionists have further explored our reality in the context of discrimination, socially and politically, and have argued that our so called universal morals and ideals have merely contributed to the creation of "inequality and oppression in the name of progress and marginalizes and disempowers individuals and groups" (Robbins, 2012, p. 329).

Michel Foucault is perhaps the most relevant social constructionist philosopher in this context. His work specifically examined the power structures of the helping professions and the science and medical information circulating in these fields. Foucault argues that social service professionals are participants in a network of social control. The "truths" by which these social service professionals operate are not truths or absolutes but "simply an effect of the rules of...power relations that create and constitute a particular form of life" (Robbins, 2012, p. 331). Power relations, according to Foucault, in communities and in our society, are not based on intelligence or aggression, but are rather random structures that perpetuate "regimes of truth" (Robbins, 2012, p. 332).

These professionals enhance what Foucault termed "biopower" which is a confine that strives to normalize the population and circulate ideas about health. Foucault was:

> especially critical about the way in which scientific knowledge is used to exert power and social control over people…the growth of the human sciences has led to a system of domination in which professionals…oppress the mentally ill, the prisoners, and society as a whole. (Robbins, 2012, p. 331)

Foucault believed that therapists and counselors only help clients to search for flaws that explain why they do not fit into the status quo. Clients are encouraged to search for or even create problems within themselves, and then they must suppress and manage these problems to fit into societal norms. In this way, Foucault supported those individuals who refused to conform or become part of the social machine that Foucault identified, allowing for the "rediscovery of particular, fragmented, subjugated, local knowledge or understanding" (Robbins, 2012, p. 332).

Politically, Foucault viewed both community level and worldwide governmental organizations as agencies of power that instituted agendas and programs based on self and institutional interest. These programs created our modern day civilization, specifically those societies formed after the fall of sovereign power. For Foucault, the move from sovereign power to his idea of biopower is the essential example of his theories on subjective reality, in which people become able to represent themselves:

> Members of a population are no longer held to be merely territorially bound participants obliged to submit to their sovereign ruler, but vital beings with their own habits, norms, and everyday practices. This means they are capable of working on themselves in their self-interest to improve and optimize their lives and life chances, and in turn, enable biopolitical interventions by various authorities to work. (Raman & Tutton, 2010, p. 714)

Our modern day democracy is the platform on which the possibility of creating our own subjective reality was born.

Marijuana Law and Cannabis Use Disorder as a Social Construction

Instinctually, understanding marijuana as an ancient substance used for recreational, medical and agricultural purposes, it is almost impossible to view the substance as an illegal, addictive drug today. The first evidence that cannabis use disorder is a social construction, is the initial publication of the DSM in 1952, which occurred in the same year that the Boggs Act was passed (Sloman, 1979, p. 189). This observation directly supports the ideas of social constructionism, as we see both the medical world and political world move to define, criminalize and medicalize a substance with no scientific evidence and with the subjective moral and professional motivations of community leaders.

Conrad and Schneider specifically address the relationship between law, medicine and socially acceptable behavior. In their 1992 work *Deviance and Medicalization: From Badness to Sickness*, the authors explore the history of the social constructions that have defined our medical knowledge and our ideas about criminal behavior. Conrad and Schneider too view morality as a social construction and acknowledge the institutions of power that define our morality. They explain that:

> Morality becomes the product of certain people making claims based on their own particular interests, values, and views of the world. Those who have comparatively more power in a society are typically more able to create and impose their rules and sanctions on the less powerful. (Conrad & Schneider, 1992, p. 2)

When we reflect on the role of Harry Anslinger, the first commissioner of the Federal Bureau of Narcotics, we recall one powerful individual who pursued his own political ideas based on his personal religious and moral views, despite consistent research that denied his statements about the dangers of marijuana. Anslinger, the motivation behind the first 30 years of U.S. drug laws, used ideas about drugs and low morals to convince a terrified public to join his campaign.

> Anslinger repeatedly rejected clinical analyses that concluded marijuana did not induce violent behavior or lead to the use of more addictive drugs...Anslinger successively tied marijuana use to

jazz, which he despised due to the prevalence of African-American musicians, WW II (arguing that the Japanese used the plant to sap the will of American prisoners), and, later, the Cold War (where Communists took the role previously occupied by the Japanese)…Several observers conclude that the "marijuana crisis" was essentially manufactured by the FBN (Galliher and Walker 1977), an example of what Goode and Ben-Yehuda (1994) label an "elite-engineered" moral panic…(Warf, 2014, p. 429-43)

Anslinger was successful in not only passing decades of legislation but encouraging the public's negative interpretation of marijuana through propaganda and sensationalized journalism.

Returning to Conrad and Schneider, we can attempt to unravel the more complex idea of the medicalization of cannabis use disorder. Now that powerful institutions and individuals have defined society's moral views of marijuana use as a degenerative, violent, and savage life style, and laws have been passed regulating these morals, all that is left is the medicalization of the drug as a substance abuse disorder. Despite the fact that hundreds of years of research, including ancient writings and artifacts, support the harmlessness of the drug, we now live in a society which insists on the harmful, addictive and debilitating nature of marijuana. How did this happen?

Conrad and Schneider explore this question. They concur that illness is just as much a social construction as politics. "The high degree of consensus on what 'objectively' is disease is not independent of the social consensus that constructs these 'facts' and renders them 'important'" (Conrad & Schneider, p. 31). They go on to remind us of the special relationship that illness and illegal activity share, and that "When such medical designations are applied to deviant behavior, they are related directly and intimately to the moral order of society" (Conrad & Schneider, p. 35). Medicalization for Conrad and Schneider:

> consists of defining a problem in medical terms, using medical language to describe a problem, adopting a medical framework to understand a problem, or using a medical intervention to "treat" it. This is a sociocultural process that may or may not involve the

medical profession. Medicalization occurs when a medical frame or definition has been applied to understand or manage a problem; (Conrad 1992: 211). (London, 2009, p. 8)

Because marijuana law forced marijuana use to become an illegal behavior, room was created for the medicalization of marijuana use: "Before a conduct or action receives a medical term or a medical definition, it must first exist under the definition of deviant" (London, 2009, p. 9). This has been accomplished over the last 100 years through marijuana legislation. Then, medical discoveries "stak[e] claims…on social space (i.e., conceptual, institutional, and/or interactional turf previously not held by the medical profession)" (London, p. 9) and the already criminalized use of marijuana becomes medicalized. "By defining the deviant as sick, Conrad and Schneider affirm, the judgment of immorality is kept. (Conrad and Schneider, 1992: 271-2)" (London, p. 11). Cannabis use disorder is born.

Marijuana use is not an experience in a vacuum and our modern day society is quick to forget the historical significance of almost everything, including cannabis. Cannabis use disorder is not a substance abuse disorder but instead, a political and social construct designed to constantly engage the majority on ideas of morality, compulsivity, racism, and science. Beginning with the regulation of the drug and then the criminalization of marijuana, our society has predictably moved to medicalize its use, as it has with other socially immoral lifestyles including homosexuality.

Yet social constructionism has its critics, and as social workers, we are still interested in treating clients who smoke marijuana. For the helping professions, a social constructionist perspective is difficult because in all essence, it presumes that individuals do not truly have problems, but instead are clashing against the status quo. This is a difficult place to be for social workers, and as the political arena becomes more entranced with marijuana legalization and scientific evidence, it is in our best interest to understand both sides of the argument.

Conrad and Schneider are open to this criticism. A major concern of social constructionist perspective is that while theorists elaborate on the lack of ultimate truth, they are simultaneously preaching their own reality and their personal perspective. More eloquently put, this may be referred to as "'ontological gerrymandering': a practice of argument and writing in which the analyst or observer is allowed a privileged place from which to see 'truth' and against which participants '--constructors'--constructing activities stand out in relief (see also Pollner, 1991, Schneider, 1992)" (Conrad & Schneider, 1992, p. 279-280). Conrad and Schneider call for a more "reflexive" social constructionist perspective which applies social constructionist arguments to social constructionists' theories (Conrad & Schneider, 1992, p. 279).

Other criticisms observe social constructionism as a "paradoxical" (Robbins, 2012, p. 332) view of the human self, as individuals are a product of historical events but at the same time able to create and re-create themselves based on their subjective reality. How can one be tied to history, formed and manipulated by it, but also free to evolve in his or her current subjective reality? Some critics suggest that postmodern thought, including social constructionism, includes theories that remove individuals from the responsibilities of social and moral expectations, allowing for ultimate freedom of existence. This idea denies our human connectedness (Robbins, 2012, p. 332). "As a result, critics worry that postmodernism encourages passivity or cynicism and thereby accelerates social atomization and personal malaise" (Robbins, 2012, p. 333).

Those who oppose social constructionism would advance several arguments, as follows: that marijuana law and cannabis use disorder are based in universal scientific fact, and that these rules and regulations are in place to protect the common good. Research proves that marijuana is a harmful and addictive substance that leads to harder drugs, criminal behavior, and a waste of life. The ancient use of marijuana occurred in societies in which modern crime was nonexistent, and the intensity of the drug was much lower. Therefore the historical use of the drug is not relevant as the modern day phenomena of urban environments and mass production create serious social problems. Cannabis use disorder is a serious substance use disorder that affects many people, including young people, and its treatment is necessary for the health and happiness of our communities (http://www.drugabuse.gov/publications/drugfacts/marijuana).

Despite this school of thought, social constructionism does not rely on the scientific evidence of marijuana's harmful or helpful chemical properties. Additionally, for as many scientific studies that conclude that marijuana is harmful, there are scientific studies that prove that it is helpful. Perhaps this is the best example of why cannabis use disorder is a social construction—there is no universal truth, only what individuals choose to believe. Social constructionism aims to explore the historical development of a complex web of power that results in the legal, medical, and political composition of our world, which is not explained by science or universal morals or truths. In this way, marijuana law and cannabis use disorder are social constructions.

Implications for Social Work

Social workers must engage with clients who have been diagnosed with cannabis use disorder or have been legally sanctioned for its use. Social constructionist theory encourages the use of the strengths perspective with these clients. The strengths perspectives:

> emphasize discovery and expansion of clients' and communities' strengths, capacities, and resources; recognition of challenges and oppression without defining people merely as victims, sick pathological, or problematic; collaborative helping relationships; focus on solutions and creative possibilities; encouraging clients to narrate and reconstruct their life stories; and deconstructing taken-for-granted assumptions about the nature of people and societies. (Robbins, 2012, p. 337)

Clients have the ability to grow and change and experience their lives as they wish and as is best for their development. As Foucault would acknowledge later in his life, individuals and clients are works of art, with the ability to change and evolve (Robbins, 2012, p. 332).

A client with a diagnosis of cannabis use disorder may strongly desire to become abstinent, in which case social workers may be relieved to work with a client who is ceasing illegal activity. Alternatively, clients who have been diagnosed with cannabis use disorder may have no interest in becoming abstinent for the exact same reasons that social constructionists view marijuana law and cannabis use disorder as faulty—and then, social workers are in a pickle. However, social workers can acknowledge and honor social constructionism by practicing strengths-based theories which allow the client to safely and healthily define their own reality and growth, with the help of a caring and supportive social worker.

Social workers have the unique experience of working in a field which can alter the subjective experience, through political and community engagement. But the fight is long and hard, and many clients suffer in the meantime in our current reality. While Foucault viewed social workers as cogs in the machine of our controlled existence, I would argue that our contemporary training works diligently to prevent robotic and inflexible professionals whose primary goal is to fit clients into the status quo. As social workers strive to become professionals who promote freedom and individualism, we are simultaneously at the mercy of slowly changing laws and the even slower maturing DSM.

Social workers walk a tightrope with clients who participate in activity that is currently deemed illegal, such as smoking marijuana, but which we identify as a result of social constructionism. It is our job to protect the client and also promote their agency and so it is my hope that social workers are coming to terms with the social constructionism of our tightly wound society, recognizing their role in it, and working to free themselves and their patients, in a socially responsible way, from the confines of our oppressed reality.

References

American Psychiatric Association., & American Psychiatric Association. (2013). *Diagnostic and statistical manual of mental disorders: DSM-5*. Washington, D.C: American Psychiatric Association.

Conrad, P., & Schneider, J. W. (1992). *Deviance and medicalization: From badness to sickness : with a new afterword by the authors*. Philadelphia: Temple University Press.

London, J. M. (2009). *How the use of marijuana was criminalized and medicalized, 1906-2004: A Foucaultian history of legislation in America*. Lewiston, N.Y: Edwin Mellen Press.

McNamee, S., & Gergen, K. J. (1992). *Therapy as social construction*. London: Sage Publications.

Robbins, S. P., Chatterjee, P., & Canda, E. R. (2012). *Contemporary human behavior theory: A critical perspective for social work*. Boston: Allyn & Bacon.

Sloman, L. (1979). Reefer madness: T*he history of marijuana in America*. Indianapolis: Bobbs-Merrill.

Stewart, A., Morrow-Cribbs, B., & Rosen, J. (2009). *Wicked plants: The weed that killed Lincoln's mother & other botanical atrocities*. Algonquin Books.

Warf, B. (2014). High Points: An Historical Geography of Cannabis. *Geographical Review*, 104(4), 414-438. doi:10.1111/j.1931-0846.2014.12038.x

Witkin, S. L. (2012). *Social construction and social work practice: Interpretations and in-novations*. New York: Columbia University Press.

Escaping the Unseen: Conduct Disorder in a New Light

Taha Zaffar

First Light

We live in a moment of history where change is so speeded up that we begin to see the present only when it is already disappearing. —R.D. Laing (1967/1990)

Rather than attempt an exhaustive inquiry into Foucault's texts, *The Birth of the Clinic* (1963) and *Discipline & Punish* (1975), the present critical consideration of the DSM-5 diagnostic criteria for "Conduct Disorder" argues instead for the application of two seminal ideas: the "medical gaze" and "panopticism." Discussed separately, and then in convergence, these Foucauldian concepts shed frightening light on the contemporary use and understanding of "Conduct Disorder."

Published and translated during the prominence of the post-structuralism movement, a post-hoc American label for the collection of works produced by 20th century French thinkers, continental philosophers, and critical theorists, Foucault's works were part of a larger corpus challenging extant literary and philosophical paradigms. While himself rejecting the label of post-structuralist, Foucault and his contemporaries marked an intellectual disillusionment with structuralism, the conceptualization that understanding human culture emerges from uncovering the structures (constituted by interrelations) undergirding human perception, sensation, and action. Conceptually borrowing heavily from the structural linguistics of de Sassure, the movement faced charges, from both linguistics and philosophy, as unduly deterministic in its understanding of human behavior, (Deleuze, 1953).

Conversely, post-structuralism—particularly Foucault's genealogical take on the social sciences—undercuts the very notion of a single, unified, absolute understanding of human culture. Instead, he delves into the historical as a construct subject to the multiplicity of perspectives, identities, and positions of viewers. More stridently, as Sarrup (1993) notes, "Genealogy...is a form of critique. It rejects the origin in favour of a conception of historical beginnings as lowly, complex, and contingent," (p. 59). This conception of history both explicitly and implicitly challenges contemporary assumptions of history's "irrationality." The "technology of power," specific mechanisms of structural concontrol, of the past—when laid bare—questions the degree to which current structures exist as bastions of rationality from an irrational past, or, rather, as Foucault would understand them, institutions developed and versed in specific techniques of power (Sarrup, 1993, p. 82).

It is in laying bare these mechanisms of power that the discussion turns to the DSM-5's conceptualization of "Conduct Disorder." First, however, it is cogent to note the degree to which "Conduct Disorder" pervades the very institution Foucault critiques: the prison. While the DSM-5 pallidly notes that, for the general population, "One-year population prevalence estimates range from 2% to more than 10%, with a median of 4%," the prevalence in juvenile detention centers is alarming (American Psychiatric Association, 2013, p. 473). Meta-analysis of surveys examining the preponderance of mental health disorders among adolescents in juvenile correctional facilities suggests a rate of 52.8% (95% confidence interval [CI] 40.9%-64.7%) for boys and 52.8% (95% confidence interval [CI] 32.4%-73.2%) for girls (Fazel, Doll, & Långström, 2008). A targeted study, The Northwestern Juvenile Project, *randomly sampled* male and female youth arrested and detained at point of entry at Cook County Juvenile Detention Center over a period of two and half years. Findings suggest, "…that nearly two-thirds of males and nearly three-quarters of females met the diagnostic criteria for one or more of the disorders listed." (Teplin, Abram, McClelland, Mericle, Duncan, & Washburn, 2006, p. 2). Prevalence for "Conduct Disorder" in particular stands at 37.8% of males and 40.6% of females (n=1,170 and 656, respectively; Teplin, et. al., 2006, p. 7). These findings are not only corroborated by a Florida study of children placed in out-of-home care (n=5,720), but augmented by the research's conclusion that, "…conduct disorder was the strongest predictor of children's subsequent involvement with the juvenile justice system." (Yampolskaya & Chuang, 2012, p. 591).

Its innocuous notation in DSM-5 notwithstanding, the prevalence of the disorder within juvenile detention centers raises disquieting questions: To what extent are detainees screened for this disorder (given the burgeoning body of data)? Does the conceptualization of this disorder predispose individuals to detention facilities? And, who or what benefits from the cyclic relationship between "Conduct Disorder" and recidivism? Efforts to address these questions lend themselves to a Foucauldian analysis, especially insofar as scrutiny is concerned with exploring the unique diagnostic criteria of "Conduct Disorder" in DSM-5 as an insidious mechanism of social control.

Prior to analysis, it is worth including the constellation of "symptoms" comprising "Conduct Disorder":

Table 1: Diagnostic Criteria for DSM-5

A.   A repetitive and persistent pattern of behavior in which the basic rights o others or major age-appropriate societal norms or rules are violated, as manifested by the presence of at least three of the following 15 criteria ir the past 12 months from any of the categories below, with at least one cri terion present in the past 6 months:

    Aggression to People & Animals
1.   Often bullies, threatens, or intimidates others.
2.   Often initiates physical fights.
3.   Has used a weapon that can cause serious physical harm to other (e.g., a bat, brick, broken bottle, knife, gun).
4.   Has been physically cruel to people
5.   Has been physically cruel to animals.
6.   Has stolen while confronting a victim (e.g., mugging, purse snatching, extortion, armed robbery).
7.   Has forced someone into sexual activity

    Destruction of Property
1.   Has deliberately engaged in fire setting with the intention of caus ing serious damage.
2.   Has deliberately destroyed others' property (other than by fire se ting).

    Deceitfulness or Theft
1.   Has broken into someone else's house, building, or car.
2.   Often lies to obtain goods or favors or to avoid obligations (i.e., "cons" others).
3.   Has stolen items of nontrivial value without confronting a victim (e.g., shoplifting,

    Serious Violation of Rules

1. Often stays out at night despite parental prohibitions, beginning before age 13 years. 2. Has run away from home overnight at least twice while living in the parental or parental
B. The disturbance in behavior causes clinically significant impairment in social, academic, or occupational functioning. C. If the individual is age 18 years or older, criteria are not met for antisocial personality disorder.

(Adapted from DSM-5, American *Psychiatric Association, 2013)*

Medical Gaze

Ever-present, and perhaps inescapable to those fortunate enough to have access to healthcare is the interaction between provider and patient. Given the disparate understanding of health processes between professionals and their patients, a degree of deference, willing subjugation to the "expertise" of the professional, contours the interactions. While the biomedical sciences, calling for years of specialized study to understand physiological systems and mechanisms, root themselves in a modernist expert-layperson configuration, the profession of Social Work arguably exists in a dialectic between modernism and postmodernism, (Gray, 2011, p. 9). Social workers balance their "expertise" against the lived experiences and subjective understandings of their clients. Doctors and nurses are not concerned with their patient's subjective understanding of cancer to diagnose and treat it. The risk inherent in both professions however, emerges as a reductionist, phenomenological understanding of an individual. Foucault, wary of biological reductionism in particular, considers the implication of what he called the "medical gaze."

Situated, and then genealogically constructed by Foucault in the discourse of the then emergent empiricism of medical sciences in the 18th and 19th centuries, the "medical gaze" concerns itself with, "the decipherment of disease in its specific characteristics...based on a subtle form of perception that must take account of each particular equilibrium (Foucault, 1963/2003, p. 15). These "equilibriums" constitute the multiplicity of qualitative and quantitative factors (e.g. body fluids, organ functioning, etc.) which equate to health. Yet, focus on these elements leaves professionals in a "...paradoxical position. If one wishes to know the illness from which he is suffering, one must subtract the individual, with his particular qualities." (Foucault, 1963/2003, p. 15).

Excising the individual from the provider-patient dyad results in a closed loop in which the "medical gaze," "…is controlled only by itself; in sovereign fashion it distributes to daily experience the knowledge that it has borrowed from afar and of which it has made itself both the point of concentration and the centre of diffusion," (Foucault, 1963/2003, p. 35). The legitimacy and *power* of the "medical gaze" is therefore derived from and propagated by the "medical gaze" itself. And, "no longer [is] the gaze of any observer, but that of a doctor supported and justified by an institution, that of a doctor endowed with the power of decision and intervention," (Foucault, 1963/2003, p. 109). Thus, the "medical gaze" institutionalizes, as not only a putative mechanism of treatment, but also a mechanism of power and control over the bodies of the patient.

At the intersection of "Conduct Disorder" and the "medical gaze," the patient finds him or herself subject to a distinctly phenomenological appropriation of his or her behaviors by institutions of power and control. Briefly consider the diagnostic criteria for "Conduct Disorder." While DSM-5 acknowledges that, "disruptive, impulse-control, and conduct disorders include conditions involving problems in the self-control of emotions and behaviors," it inconsistently applies these considerations across diagnoses, (American Psychiatric Association, 2013, p. 461). "Oppositional Defiant Disorder" for example includes potentially *subjective* diagnostic criteria associated with angry/irritable mood, (American Psychiatric Association, 2013, p. 462). "Intermittent Explosive Disorder" criteria discuss the degree to which an individual's response is commensurate with "precipitating psychosocial stressors," (American Psychiatric Association, 2013, p. 466). Conversely, "Conduct Disorder" criteria read more like a premier catalog of criminal offenses. Outside of post-hoc disorder specifications, **no mention** of emotional regulation, affect, or presence of psychosocial stressors exists.

Patients or clients under consideration for a diagnosis of "Conduct Disorder" are therefore subject to the "medical gaze." Given the differentially distinct phenomenology of "Conduct Disorder" within DSM-5, poststructuralist analysis in Foucauldian tradition suggests that the conceptualization of "Conduct Disorder" unduly and exceptionally dehumanizes individuals through reducing them to their "crimes." Worse yet, those observers, purportedly qualified to make diagnostic decisions, institutionalized and institutionalizing have power over those they deem "disordered." Consider again the distressing prevalence rates of "Conduct Disorder" in juvenile detention facilities, the high rates of recidivism, and the cyclicality of diagnosis and detention, diagnosis and detention. An individual arguing against his or her "Conduct Disorder" diagnosis in opposition to authorities, the eyes behind the "medical gaze," fulfills an associated feature supporting diagnosis

Panopticism

Arguably, one of Foucault's most enduring concepts, "panopticism" (literally a system or ideology of all-seeing) continues his genealogical critique of ubiquitous institutions. Incited by Jeremy Bentham's prison design in the late 18th century, Foucault came to understand the Panopticon (see Figure 1) as both a symbol of and mechanism for social control, (Foucault, 1975/1979). The Panopticon houses a central tower in a vast room replete with cells. A guard standing in the tower has full, 360-degree vision of each individual cell. Despite this, it is impossible for the guard to view every cell at any given time; his or her view is limited to his or her field of vision. However, there is no way for the prisoner in any given cell to know where the guard is looking. As such, the Panopticon creates a repressive space where "...the inmate must never know whether he is being looked at any one moment; but he must be sure that he may always be so." (Foucault, 1975/1979, p. 201).

Although architecturally imposing, the Panopticon's dehumanization truly emerges from its capacity for psychological subjugation. Because a prisoner does not know when he or she is the subject of observation, he or she assumes there is a chance that he or she is *always* the subject of observation. In internalizing this perpetual surveillance, individuals themselves become the very instruments of their control. As Foucault (1975/1979) unflinchingly points out, "He who is subjected to a field of visibility, and who knows it, assumes responsibility for the constraints of power; he makes them play spontaneously upon himself; he inscribes in himself the power relation in which he simultaneously plays both roles; he becomes the principle of his own subjection." (p. 202-203).

"Panopticism" therefore rises out of the social, theoretical application of panoptic configurations and purposes to extant social institutions. In *Discipline*

& Punish, Foucault primarily traces the advent of the prison. Much in the same way that he genealogically constructs the clinic and its tool of control, the "medical gaze," Foucault delves into the birth of the modern prison by charting its institutionalization and the institutionalization (quite literally) of its mechanism of control: the Panopticon. Since then, "panopticism" has emerged as a viable critique for other institutions as well. Ericson & Haggerty (2006) consider market surveillance vis-à-vis the proliferation of consumer data mining and the potential for algorithms to predict behavior. St-Pierre & Holmes (2008) examined the "institutional violence" of what they argue to be the panoptic discipline of nurses. Graham & Neu (2004) go so far as to charge the institution of standardized testing with manufacturing "governable persons" through the panoptic implementation of test scores. Across these settings, mechanisms for penetrating observation, combined with a constructed perception of pervasive observation subjugate those at the mercy of the institution (in this case consumers, workers, and students, respectively). Unfortunately, the same power relations appear when a Foucauldian, panoptic hermeneutic encounters "Conduct Disorder".

Recall previous discussion of the "medical gaze." Consider again the striking degree to which diagnostic criteria for "Conduct Disorder" hinge on phenomenological observations. Underpinning these observations remains an inescapability of phenomenology: its reliance on the "other" to observe. It is worth pausing then when DSM-5 explicitly warns that, "Because individuals with conduct disorder are likely to minimize their conduct problems, the clinician often *must rely on additional informants*," (American Psychiatric Association, 2013, p. 472, italics added). This is problematic within a panoptic framework for several reasons, chief among them being a subversion of clinician-client relations. Diagnoses develop discursively between the clinician and client via the vehicle of language (Ishibashi, 2005). When the client's power in shaping his or her diagnosis via words is suspect because he or she is "minimizing their conduct problems", the dehumanizing "medical gaze" of the clinician stands alone. Because of this, and an insistence on "additional informants," the current diagnostic criteria for "Conduct Disorder" fulfill a panoptic prophecy of unyielding surveillance. This is not so far-flung. Schools in the UK employ closed-circuit television (CCTV), ostensibly for security (Hope, 2009).

In sum, the entry for "Conduct Disorder" in DSM-5 suggests, "it is necessary to consider the reports by others" and cautions, "...individuals with conduct disorder with [limited prosocial emotions] may not readily admit to traits in a clinical interview," (American Psychiatric Association, 2013, p.470; 472). An individual suspected of "Conduct Disorder" therefore takes the stand as a hostile witness in their own defense, even when DSM-5 readily admits that, "It not uncommon for individuals with conduct disorder to come into contact with the criminal justice system for engaging in illegal behavior," (American Psychiatric Association, 2013, p. 474). Thus, bringing to bear Foucault's "panopticism" on "Conduct Disorder" reveals a set of criteria consonant with paradigms of power, institutionalized (as in the case of the school) to repress subjects by identifying them as "disordered," criminals, or not. Foucault (1975/1979) leaves readers with the following charge, "The judges of normality are present everywhere. We are in the society of the teacher-judge, the doctor-judge, the educator-judge, the *'social-worker'-judge*; it is on them that the universal reign of the normative is based; and each individual, wherever he may find himself, subjects to it his body, his gestures, his behaviour, his aptitudes, his achievements" (p. 304, italics added).

From Prison Photography

Figure 1: Interior of Penitentiary at Stateville Prison, Joliet, IL illustrating the "Panopticon"

Convergence

Is it surprising that prisons resemble factories, schools, barracks, hospitals, which all re-semble prisons?" (Foucault1975/1979, p. 228)

Discussion heretofore has primarily concerned itself with the "medical gaze" in clinical settings and panoptic considerations for prisons and schools. Following these threads a post-structuralist, Foucauldian reading of "Conduct Disorder" in DSM-5 finds the disorder egregiously discounting the psyche and personhood of an individual. Where other disruptive, impulse-control, and conduct disorders offer diagnostic considerations of emotional tension or affect, "Conduct Disorder" merely provides a list of criminal activity. Implemented as such, the diagnostic criteria for "Conduct Disorder" reduce the individual, via the "medical gaze," to the sum of their phenomenological presentation. Simultaneously, DSM-5 calls for the all-seeing eyes of others, the clinician included, to determine the veracity of the diagnosis. By casting doubt on the behavior of clients, appropriating their actions as supporting features for diagnosis, the DSM-5 conceptualizes individuals with "Conduct Disorder" as prisoners in the Panopticon.

Unfortunately, contemporary use of the "medical gaze" and "panopticism" converge to create total institutions of social control. To illustrate, consider the institution of the school. Envisioned as public institutions to provide educational service to as much of the population as possible, schools must accommodate a growing diversity of learners. Because of this, mandated provision of specialized education to those outside of an agreed-upon norm exists (Kaufman & Hallahan, 2005). Except if, a child is not "appropriately" emotionally disturbed. The Individuals with Disabilities Education Act (2004) which details the preconditions mandated for special education, includes a section on "emotionally disturbed" students (see Appendix A). Interestingly, "One of the most widely criticized and controversial aspects of the definition are its exclusion of children who are socially maladjusted but not emotionally disturbed," (Hallahan & Kauffman, 2014, p. 184). Even more interesting, "Some states and localities have started to interpret social maladjustment as conduct disorder..." (Hallahan & Kauffman, 2014, p. 184). In other words, "Conduct Disorder", which is in the Diagnostic and Statistical Manual for Psychiatric Disorders, in some places, does not qualify a child for special education services under the Individuals with Disabilities Act.

And for those students who violate rules because of their "Conduct Disorder"? They contribute to the exceptionally alarming rate of individuals funneled from the school system into the modern prison system of discipline and control. Disproportionately impacting students of color, the school-to-prison pipeline, "involves a set of interactions between and among children, youth, their families, school personnel, other service providers, and gatekeepers of outcomes ," which, "contribute to a cycle of negative encounters," (Osher, Coggshall, Colombi, Woodruff, Francois, & Osher, 2012, p. 284). Chief among these negative encounters is entry into the prison system. While restorative justice models (which necessarily involve dismantling power hierarchies) show some promise, (Schiff, 2013), the line continues to flow. Concordant with Foucault's genealogical critique of institutions, a critical perspective on the school-to-prison pipeline suggests it operates as a "net of social control", (Irby, 2014). The present paper goes so far as to suggest that "Conduct Disorder" is a trawl of this net—a mechanism for social control.

Looking back, recall that diagnosis of "Conduct Disorder" essentially requires a phenomenological presentation of criminal activity for diagnosis. The high rate of incarcerated youth who meet the criteria for its diagnosis alone gives pause. Yet, down to the diagnostic language, "Conduct Disorder" dehumanizes, disenfranchises and disempowers youth by subjecting them to the "medical gaze" of the normative judges, encouraging their observation, and leaving them at the mercy of panoptic power structures.

Ultimately, the diagnostic understanding of "Conduct Disorder," particularly when viewed in the Foucauldian tradition, hints at a total institution of social control emergent at the convergence of prisons and mental health treatment. A detailed, joint report by the Treatment Advocacy Center and National Sheriffs' Association charts deinstutionalization, the emptying and closing of state mental hospitals, and the subsequent proliferation of mentally ill inmates in prison systems (Torrey, Kennard, Eslinger, Lamb, & Pavle, 2010). Key findings from the report point out that, "there are now more than three times more seriously mentally ill persons in jails and prisons than in hospitals," and that, quite frankly, "America's jails and prisons have become our new mental hospitals,"(Torrey, et al., 2010, p. 1). This marks an escalation from numbers in 2005 when the U.S. Bureau of Justice found that, "more than half of all prison and jail inmates had a mental health problem," (James & Glaze, 2006). Foucault would find the alignment of the prison and mental health system troubling. In both *Birth of the Clinic* and *Discipline & Punish*, he critiques the institution, wary of its potential to subjugate bodies through its mechanisms of power. Examining "Conduct Disorder" vis-à-vis its diagnostic language and its according implications for dehumanization reveals it as mechanism of institutional power. Disquietingly, the prison, subsuming the role of state mental health agencies—while intimately connected to the school system via pipeline—realizes Foucault's worst fears.

References

American Psychiatric Association. (2013). *Diagnostic and statistical manual of mental disorders* (5th ed.). Washington, DC: American Psychiatric Association.

Deleuze, G. (1953). How do we recognize structuralism?. *Desert Islands and Other Texts, 1974,* 170-192.

Ericson, R. V., & Haggerty, K. D. (Eds.). (2006). *The new politics of surveillance and visibility.* University of Toronto Press.

Fazel, S., Doll, H., & Långström, N. (2008). Mental disorders among adolescents in juvenile detention and correctional facilities: a systematic review and metaregression analysis of 25 surveys. *Journal of the American Academy of Child & Adolescent Psychiatry, 47*(9), 1010-1019.

Foucault, M. (1979). *Discipline & punish: The birth of the prison.* (A. Sheridan, Trans.).New York, NY: Great Britain. (Original work published 1975).

Foucault, M. (2003). *The birth of the clinic: An archaeology of medical perception.* (A. Sheridan, Trans.). New York, NY: London. (Original work published 1963).

Gray, M. (2011). A Critique of the strengths perspective. Families in Society 92(1), 5-11.

Graham, C., & Neu, D. (2004). Standardized testing and the construction of governable persons. *Curriculum Studies,* (36).

Hallahan, D. & Kauffman, J. (2014) *Exceptional Learners: Introduction to Special Education* (13th ed.). New York: Pearson Education.

Hope, A. (2009). CCTV, school surveillance and social control. *British Educational Research Journal, 35*(6), 891-907.

Individuals With Disabilities Education Act, 20 U.S.C. § 1400 (2004).

Irby, D. J. (2014). Trouble at school: Understanding school discipline systems as nets of social control. *Equity & Excellence in Education, 47*(4), 513-530.

Ishibashi, N. (2005). Barrier or bridge? The language of diagnosis in clinical social work. Smith College   Studies in Social Work, 75 (1), 65-80.

James, D. J., & Glaze, L. E. (2006). *Mental health problems of prison and jail inmates.* Washington, DC: US Department of Justice, Office of Justice Programs, Bureau of Justice Statistics.

Kauffman, J. & Hallahan, D. (2005) *Special Education: What Is It and Why Do We Need It.* New York: Pearson Education

Laing, R. D. (1990). *The politics of experience and the bird of paradise.* Penguin UK. (Original work   published 1967).

Osher, D., Coggshall, J., Colombi, G., Woodruff, D., Francois, S., & Osher, T. (2012). Building school and teacher capacity to eliminate the school-to-prison pipeline. *Teacher Education and Special Education: The Journal of the Teacher Education Division of the Council for Exceptional Children,* 35(4), 284-295.

Sarup, M. (1993). *An introductory guide to post-structuralism and postmodernism.* Pearson Education.

Schiff, M. (2013, January). Dignity, disparity and desistance: Effective restorative justice strategies to plug the "school-to-prison pipeline.". In *Center for Civil Rights Remedies National Conference.   Closing the School to Research Gap: Research to Remedies Conference. Washington, DC.*

St-Pierre, I. & Holmes, D. (2008). Managing nurses through disciplinary power: a Foucauldian analysis   of workplace violence. *Journal of Nursing Management, 16*(3), 352-359.

Torrey, E. F., Kennard, A. D., Eslinger, D., Lamb, R., Pavle, J. (2010). *More mentally ill persons are in jails and prisons than hospitals: A survey of the states.* Arlington, VA: Treatment Advocacy Center, 2010.

Teplin, L., Abram, K., McClelland, G., Mericle, A., Dulcan, M., and Washburn, J. (2006).*Psychiatric Disorders of Youth in Detention*. Office of Juvenile Justice and Delinquency Prevention Juvenile Justice Bulletin, NCJ 210331, Washington, DC

[Untitled photograph of guard tower at Stateville Penitentiary]. Retrieved from http://prisonphotography.org/2010/page/17/

Appendix A

i. Emotional disturbance means a condition exhibiting one or more of the following characteristics over a long period of time and to a marked degree that adversely affects a child's educational performance:

    a. An inability to learn that cannot be explained by intellectual, sensory, or health factors.

    b. An inability to build or maintain satisfactory interpersonal relationships with peers and teachers.

    c. Inappropriate types of behavior or feelings under normal circumstances.

    d. A general pervasive mood of unhappiness or depression.

    e. A tendency to develop physical symptoms or fears associated with personal or school problems.

i. Emotional disturbance includes schizophrenia. The term does not apply to children who are socially maladjusted, unless it is determined that they have an emotional disturbance under paragraph (c)(4)(i) of this section.

From IDEA (2004)

Depression and Society

Lauren Rojas

Everyone experiences sadness. Whether the loss of employment or friendship, an unexpected setback, or dissatisfaction with one's place in life, sadness and unhappiness are common emotions that are oftentimes balanced out by feelings of joy and success if a person is mentally healthy. When a person is unable to recall the last time he or she felt happy, then depression has permeated his or her life. While other mental illnesses can seem obscure and unknown, even a person who has never experienced depression can comprehend what impact depression has. We cannot fend off the media's exposure to commercials for depression medication to help manage and alleviate symptoms of depression. The media is also constantly exposing us to various pressures to succeed and perform. It is also important to take a step back and examine how the impact of stressors--spoken and unspoken--placed on us by society take a toll on a person's mental health. By examining major depressive disorder from a societal standpoint through the lens of social constructivism and Erikson's psychosocial theory, depression can be addressed and understood with a more humanistic, compassionate, and holistic approach.

Society today is constantly being inundated with notions of expectations and pressures. Expectations to look a certain way, have a certain job, live in a certain home or neighborhood, have a certain amount of children are all inescapable impressions that infiltrate all aspects of our activities and interactions. It may not be a given that a person acts in a specific manner, but not doing what is silently expected of us leads us to not live up to society's standards. We begin to form our own beliefs on what identity we should create based on the social factors and influences that surround us daily.

DSM-V Major Depressive Disorder Diagnosis

Before examining depression through the lens of both theories, it is important to have a substantial understanding of depression viewed through the DSM-V and current research on the illness. Major depressive disorder is characterized by discrete episodes of at least two weeks duration involving distinct changes in affect and cognition. When diagnosing a person with major depressive disorder, it is critical to take into account the difference between general sadness and grief from a major depressive episode. The DSM-V states for a diagnosis of major depressive disorder, five or more of the following symptoms must be present during the same 2-week period and represent a change from previous functioning: depressed mood most of the day, nearly every day, marked diminished interest in all, or almost all, activities most of day, nearly every day, significant unintentional weight loss or decrease or increase in appetite, insomnia or hypersomnia nearly every day, psychomotor agitation, fatigue or loss of energy, feelings of worthlessness or guilt, diminished ability to cognitively function clearly, and recurrent thoughts of death or suicidal ideation without a plan (DSM-V, 2013).

The primary symptoms related to depression are sadness, emptiness, irritable mood, and somatic and cognitive changes which significantly affect the individual's capacity to function. Major depressive disorder can first appear at any age, but the likelihood of onset increases with puberty, and incidences appear to peak in the 20s, and first onset in late life is not uncommon. People between the ages of 18-29 are three times more likely to experience symptoms of depression than individuals over 60 years of age (DSM-V, 2013). This statistic reinforces the impact of societal expectations on us and pressures to succeed and perform that occur around a person's 20s. The illness affects the lives of 7% of people in the United States. Various treatments of depression range from medications, psychotherapy, and hospitalization and residential treatment. The most powerful stressors that lead to depression have been proven to be the death of a close relative, experiencing assault, and extreme relationship conflict. Persons diagnosed with major depressive disorder feel as though their illness seeps into their every decision, move, and relationship.

In order to receive a diagnosis of MDD, depressed mood must be present for most of the day nearly every day (DSM-V, 2013). Fatigue and sleep disturbances are highly common, and "the essential feature of a major depressive episode is a period of at least 2 weeks during which there is either depressed mood or the loss of interest or pleasure in nearly all activities" (DSM-V, 2013, 163). Things that people suffering from MDD once enjoyed no longer bring pleasure to their lives. Hobbies, activities, and relationships that brought joy to a person's life become ambivalent and unimportant. People suffering from MDD describe their mood as "depressed, sad, hopeless, discouraged, or 'down in the dumps'" (DSM-V, 2013, 163). The will to live and ability to find joy in daily life dissipates. Even the smallest task requires a substantial amount of effort (DSM-V, 2013). People with MDD suffer from a sense of worthlessness which may also include "unrealistic negative evaluations of one's worth or guilty preoccupations or ruminations over minor past failings" (DSM-V, 2013, 164). Previous events in a person's life that could potentially have been resolved come back to haunt them. They see their constant failings in every aspect of life. Suicidal ideations, thoughts of death, or suicidal attempts are common occurrences ranging from not wanting to get out of bed in the morning to thinking people would be happier if they were no longer alive to a specific suicide plan (DSM-V, 2013).

Major Depressive Disorder does not make an appearance in a person's life - it becomes their life. It plays a role in every single aspect of a person's life, and even though its mark may ebb and flow, it is never invisible. Given that major depressive disorder greatly impacts the lives of almost one out of every fourteen people in the United States, analyzing its prevalence and criteria needs to be done with a less clinical approach and more motivational and societal approach. Examining depression through a combination of psychosocial theory and social constructivism will allow the diagnosis of major depressive disorder to be seen through a modern approach and to see it as occurring because of the burdens we feel to thrive--and at times, simply to survive.

Analysis of Psychosocial Theory and Social Constructivism

Erik Erikson's ideas were highly influenced by those of Freud while emphasizing the role of society and culture on a person's natural progression and subconscious. He acknowledged "the importance of social variables such as family, community, and culture shaping the individual" (Coady & Lehmann, 2008, 120). He created the idea of a person's life consisting of eight stages to create a lifespan model within which the stages "provided an opportunity for the individual to learn new skills for progressing to the next stage" (Coady & Lehmann, 2008, 120). Within the eight stages--trust versus mistrust, autonomy versus shame, initiative versus guilt, industry versus inferiority, identity versus identity confusion, intimacy versus isolation, generativity versus stagnation, and integrity versus despair--a person is confronted with a crisis that must be resolved in order to successfully progress to the subsequent stage (Coady & Lehmann, 2008). Failing to resolve a given crisis holds a person back from the next stage until they find the resolution. For some people, this inability to discover a resolution leads to detrimental regression and remaining at a standstill in life. Especially for children who are extremely impressionable and malleable, to see another child progress while you are motionless can create a great likelihood for future difficulties following later in life.

Postmodernist philosophy posits that human experience is socially constructed. Social constructivism views the world through the lens that an individual's reality and truth is mentally constructed. Authors Coady and Lehmann state, "social constructivism focuses on the power of social interaction and culturally shaped assumptions for shaping knowledge and meaning" (Coady & Lehmann, 2008, 56). Diagnostic labeling is approached with caution and skepticism in social constructivism due to its characteristic negative, deficit meanings, and the implication that an external, knowledgeable expert can possibly narrowly classify another human being. For most constructivists, the *Diagnostic and Statistical Manual of Mental Disorders* represents an invented model of abnormality in which mental disorders are viewed as naturally occurring, objective entries. (Coady & Lehmann, 2008, 405)

Coady and Lehmann address the issues regarding the diagnoses covered in the DSM-V and the process of diagnosing. The narrow and clinical view through with the DSM-V applies mental illnesses negates the external factors and cultural aspects social constructivists believe play a large part in a person's world.

Social Constructivism can be viewed as the modern approach on Erikson's revolutionary thinking. Both theories relate the world we live in to the world we experience individually. Erikson believed that society creates tasks that we must complete in order to move on and be successful, while social constructivism embraces the concept that humans connect their own meaning to life experiences. Psychosocial Theory and Social Constructivism are interrelated as they both interplay the relationship between a person and what it means to be successful based on what society deems as expected. Without the expectations and standards society sets for us, psychosocial stages would have no validity and social constructivism would have no reasoning behind its theory.

Major Depressive Disorder Through the Lens of Erikson and Society

Society tells us that children must be involved in multiple sports and hobbies while fostering their imagination and completing school with excellent grades in order to find success later in life. We then must be successful in our career with a house and healthy relationship while also being financially sound, healthy and active, and maintaining perfect relationships with our family and friends. While we can tell ourselves that those pressures mean nothing on our life and success is intrinsic based on what an individual values, it is almost impossible to turn off the outlying noise and dismiss what we are told. We also see people around us completing these life tasks and compare our progress to theirs.

On top of all the expectations, we are thrown curveballs by life. Psychosocial factors such as stressful life circumstances and stress factors are a major concern today, and depression is seen as resulting from the interplay of personal and environmental resources, environmental stressors, and an individual's ability to cope with stressful events. In putting depression and social theories together, one can argue that societal expectations are a great contributor and benefactor to a person's likelihood to experience depression in their lifetime. No matter how strong or supported a person may be, environmental influences take a large toll on a person's mental stability.

Several pieces of research assist in connecting the topics of depression and social theories to create a cohesive relation. Authors Song, Eagle, and Mickelson (2001) research the impact of bullying in childhood and life goals and choices as an adult. They state that in examining a person's inability to discover their passion and needing to return to school later in life can be viewed through the lens of Erikson's psychosocial stages as an incomplete transition due to previous bullying between certain life tasks and stages, and those individuals can be more likely to experience depression (Song, Eagle & Mickelson, 2001). This research emphasizes the notion of pressures not only as an adult but as a child as well and how those stressors can influence decades of a person's life, if not its entirety.

Research by Brott (2005) on how a constructivist would view life roles through counseling references the "influence of an individual's occupation on his or her place in society, circle of friends and acquaintances, use of leisure time, political affiliates, interests and aspirations, and boundaries of culture" (Brott, 2005, 140). This occupational choice a person must has the ability to nurture or impair a person's entire life, and this choice can oftentimes stem from not necessarily what a person knows will make them happy but what he or she is expected to be.

Authors Shulman and Nurmi (2010) delve into understanding emerging adulthood from a goal-setting perspective. Their research shows that "individuals who report many interpersonal, school and leisure-related goals show less burnout, stress, and depression than those who report less-adaptive goals" (Shulman & Nurmi, 2010, 7). The authors also state "young adults who evidence progressive development in goals related to both work and love show low levels of depression, whereas lack of goals in both domains was associated with increased depression" (Shulman & Nurmi, 2010, 7). The societal pressures constantly being communicated can lead people, especially children and young adults, to feel so overwhelmed that they simply surrender and lose the desire to create goals, which can lead to depression through a lack of self-identity and drive. These findings help to highlight that no age frame is immune from the incessant notions of what success is and what we are expected to achieve.

Upon studying the impact of psychosocial developmental task resolution and the relation between early maladaptive schemas, Author Thimm (2009) references Erikson's research on unsuccessful task resolution leading to maladaptive schemas (Thimm, 2009). The author continues by discussing the correlation between resolution of tasks and maladaptive schemas and states "depending on successful or unsuccessful resolution of a developmental task, the individual acquires a more or less favorable ratio between positive or negative attitudes...Failing (to resolve a task) may hamper the resolution of the other developmental tasks" (Thimm, 2009, 222). In addition to research on maladaptive schemas, task resolution, and depression, Authors Harris and Curtin (2002) delve deeper into the connection between schemas research development and correlation with depressive symptoms. The authors discover a positive correlation between the score a person receives on a Schema Questionnaire and symptoms of depression. These findings help to solidify the finding that these personality traits which can be attributed to what is expected of us throughout our lives and in developing our personalities can lead to a development of depressive symptoms. When failing to resolve a developmental task, a person is stuck while the people around them progress forward. Seeing others grow and flourish while being stuck creates distress and can oftentimes make it more difficult to resolve the task because of lack of motivation.

From these various studies one can infer the impact of burdens we feel from our community and culture which can ultimately lead to distress. A child who experiences the pathology of bullying becomes unable to resolve a psychosocial task. This inability to resolve a stage paralyzes the child in his current stage while his peers continue to thrive and progress. The child realizes his peers are moving on which could either motivate the student to work harder towards a resolution or lead him to give up. While this is occurring, the child also continues to progress through his education and continues receiving constant reminders of where the child is expected to be based on the standards and assumptions society creates and notices how behind the child is compared to peers. As this progresses throughout life, the child progresses to adolescent and adult and the inability to resolve a stage early in life leads to turmoil and depression. While this instance is not applicable to every person, experiencing an unresolved psychosocial stage mixed with societal pressures can lead a person to a life of depression.

Eliminating the societal pressures of today is a fascinating, intriguing, and overwhelming thought. Society dictates so much of people's decisions and stands as a barometer for some as to what their life should look like. Without those notions of success, people will discover the freedom to create a life based on their own hopes and desires. Oftentimes we fall privy to what is expected of us over what we truly desire. Life is too short to allow external factors to wreak havoc on our reality and allow depression to overcome our lives. While there are an unlimited amount of positive life events a person can experience and positive marks to be made on society, the demands of life and what is expected of us can be too much.

A dispute to the claim that society's impact on depression is not emphasized enough in treatment could be that depression is a chemical imbalance and the stressors and pressures of society have no impact. The DSM-V acknowledges the extensive literature present on the biological impact on depression (DSM-V, 2013). Author Gabbard (2014) notes several studies which address the genetic factor while emphasizing the importance of circumstance and experiences on the development of depression. Gabbard describes one study done in 1993 in which "680 female-female twin pairs of known zygosity to determine whether an etiological model could be developed to predict major depressive disorders" (Gabbard, 2014, 219). The results of the study proved genetic factors to be substantial but underwhelming, and "the most influential predictor was the presence of recent stressful events" (Gabbard, 2014, 219). Another study conducted in 1995 hoped to uncover more regarding the causes of depression and found "sensitivity to the depression-inducing effects of stressful life events appears to be under genetic control" (DSM-V, 2014, 219). These studies continue to acknowledge the biological factors impacting depression while acknowledging the aspects out of our control that play a large part in depression's development.

To place the majority of a cause for diagnosing a person with major depressive disorder on biology and a chemical imbalance and simply accepting that as a person's fate is doing that person and society a disservice. Biological factors will always play a part; the strains of society need not always. Social workers have the opportunity to be present and involved in the movements through a macro level of activism or policy and also a micro level of working one-on-one with an individual's struggling to suppress the constant pressures. Through whichever approach is chosen, a constant reminder of the power of an individual and the strength every single person possesses needs to remain the focus of our work.

References

American Psychiatric Association. (2013). *Diagnostic and statistical manual of mental disorders* (5th ed.). Washington, DC: American Psychiatric Association.

Brott, P. (2005). A Constructivist Look at Life Roles. *The Career Development Quarterly*, *54*(2), 138-149.

Coady, N., & Lehmann, P. (2008). *Theoretical Perspectives for Direct Social Work Practice: A Generalist-Eclectic Approach. Springer Series on Social Work.* (2nd ed.). New York, New York: Springer Publishing Company LLC.

Gabbard, G. (2014). *Psychodynamic psychiatry in clinical practice* (5th ed). Washington, D.C.: American Psychiatric Press.

Harris, A., & Curtin, L. (2002). Parental Perceptions, Early Maladaptive Schemas, and Depressive Symptoms in Young Adults. *Cognitive Therapy and Research, 26*(3), 405-416.

Shulman, S., & Nurmi, J. (2010). Understanding emerging adulthood from a goal-setting perspective. *New Directions for Child and Adolescent Development,* 1-11.

Song, S., Cary, P., Eagle, J., & Mickelson, W. (2001). Psychosocial correlates in bullying and victimization: The relationship between depression, anxiety, and bully/victim status. Bullying Behavior: Current Issues, Research, and Interventions, 2, 95-121.

Thimm, J. (2009). Relationships between early maladaptive schemas and psychosocial developmental task resolution. *Clinical Psychology & Psychotherapy, 17*(3), 219-230.

Society's Acceptable Poison

Olaide Agunloye

As far as wellness is concerned, alcohol continues to generate debates. Studies of health risks related to alcohol use are often countered by studies outlining health benefits of alcohol consumption. Media promotion has not wavered, leaving the decision up to the masses. Is alcohol society's socially acceptable poison? Does it help or hurt? While pondering such queries one must not fail to acknowledge the economic benefits that have resulted from liquor utilization. According to the Distilled Spirits Council of the United States (2015), "the beverage alcohol industry is a major contributor to the economy, responsible for over $400 billion in total U.S. economic activity in 2010, generating nearly $90 billion in wages and over 3.9 million jobs for U.S. workers." All standpoints provide a different frame to analyze society's love-hate affair with alcoholism.

History

The scope of available resources and literature outlining the history of alcohol are wide-ranging and complex. The use and abuse of alcohol and its relationship to health and medicine in America generally reflects deep ideological and moral viewpoints (Stolberg, 2006). Some of the implications that have been comprised in documented literature have existed for millennia. Although useful, the amount of information that can be covered in this paper is limited. Even so, key topics will be addressed to gain a full understanding of alcohol use and alcoholism in society.

Much research suggests that alcohol has been present in human civilization for thousands of years. The Foundation for a Drug-Free World (2015) gives a brief history of alcohol, listing ancient discoveries of fermented beverages that existed in early Egyptian civilizations. Evidence places alcohol in China as early as 7000 B.C. and in India between 3000 and 2000 B.C. As time progressed, records reference a wine goddess in 2700 B.C who was worshipped by the Babylonians. Civilizations have left behind imprints of fermented recipes and drinks. As consumption became popular, issues arose. One of the earliest warnings about excessive alcohol consumption can be found in ancient Greek literature (The Foundation For A Drug-Free World, 2015).

By the sixteenth century, alcohol was used primarily for medicinal purposes (The Foundation For a Drug-Free World, 2015). Native Americans developed alcoholic beverages from crops like corn and grapes. As colonial America was established, alcohol was referenced as a source of healing. "Alcohol was considered essential for good health, warming of the body, aiding in digestion, and fortifying the constitution" (Stolberg, 2006). By the eighteenth century in Britain, cheap spirits flooded the market and reached a peak (Stolberg, 2006). Immediately following the parliament's decision to encourage the use of grain for the distilled spirits, consumption reached 18 million gallons, and alcoholism became widespread (The Foundation For a Drug-Free World, 2015). Alcohol was still used as a sedative for amputations and surgeries (Stolberg, 2006). Drunkenness, however, was considered a sin during the early Americas despite the fact that heavy alcohol consumption was a part of day-to-day life (Stolberg, 2006).

The idea of sobriety began to gain popular approval in the mid-nineteenth century. It was reported that almost all criminal activity in America was related to the abuse of alcohol in 1831 (Stolberg, 2006). By this time civilians began to view the effects of inebriation from a moral lens. The cultural shift in society's moral compass lead to the 1920 law prohibiting the manufacture, sale, import and export of intoxicating liquors in the United States. The prohibition resulted in unintended consequences, like organized crime, making criminals of many bootlegging Americans. The prohibition was short-lived, lasting only 13 years. The prohibition of alcohol was cancelled in 1933 (The Foundation For a Drug-Free World, 2015).

Current Issues

Today, an estimated 15 million Americans suffer from alcoholism and 40% of all car accident deaths in the United States involve alcohol (The Foundation For a Drug-Free World, 2015). The prevalence of alcoholism has expanded across demographics and as a result alcoholism has become a common disorder in the United States. According to the *Diagnostic and Statistical Manual of Mental Disorders* fifth edition (DSM-V), the 12-month prevalence of alcohol use disorder is estimated to be 4.6% among 12-17 year olds and 8.5%

among adults age 18 years and older. These rates continue to increase among adults age 18 years and older. Adult men reportedly have rates of 12.4%, nearly tripling the amount of adult women at 4.9% (American Psychiatric Association, 2013). Ethnically, Native American adults and Alaskan Natives have the highest rates of alcohol use disorder (American Psychiatric Association, 2013). Compared to other groups they stand at 12.1%, followed by whites at 8.9%, Hispanics (7.9%), African Americans (6.9%) and Asian Americans/Pacific Islanders (4.5%).

The history and current statistics demonstrate how alcohol has evolved. From ancient medicine to mass consumption, it appears that despite morality, alcoholism persists. This could have something to do with the associated benefits that scientists continue to proclaim. Studies have suggested that moderate drinking may improve cognitive and cardiovascular functioning while decreasing the risk of diabetes, not to mention that the new United States dietary guidelines suggest one to two drinks daily. Conversely, the dietary guidelines do stress the importance of drinking in moderation while warning against binge drinking and the associated adverse effects. For most Americans, it is easy to see the contradictions in the health care world, although it may be difficult to form an opinion on alcohol consumption.

As previously stated, alcohol is economically beneficial to society as well. This creates a global interest in the regulations that restrict alcohol related conditions. At the 2010 National Conference of State Liquor Administrators, it was observed that the majority of panelists were from the alcohol industry (Mart, 2012). It was also found that the few panelists from federal government agencies responsible for regulating the alcohol industry spoke in support of the industry instead of protecting public safety (Mart, 2012). This research suggests that alcohol regulators in the United States are more interested in the economic development that alcohol provides as opposed to the health risks that affect the public.

Diagnosis

According to the American Psychiatric Association's website, the *Diagnostic and Statistical Manual of Mental Disorders* (DSM) is the standard classification of mental disorders used by mental health professionals in the United States. The DSM entails the diagnostic classifications, the diagnostic criteria sets, and the descriptive text. The manual is one of the most popular texts used in the field of healthcare. The DSM outlines the criteria needed to identify alcohol use disorder. The fifth edition of the manual was issued in 2013, revised from the previous edition, which was released in 1994. There is a gap of almost two decades between the manuals, and the changes between the two versions are significant. One alteration in terms of terminology is that prior to the fifth edition two major disorders existed in association with alcoholism. Alcohol abuse and alcohol dependence were defined in detail in the DSM-IV. The DSM-V combines the two disorders into one. The new term to reference the diagnosis has been changed to alcohol use disorder (AUD).

The purpose of the DSM has been to improve the way that practitioners understand and diagnose mental disorders. While the DSM is widely recognized as the "go to" for diagnosis, it can also be said that it is a financial billing guide. The DSM-V is completely compatible with the HIPAA-approved International Classification of Diseases, Ninth Revision, and Clinical Modification (ICD-9-CM) coding system, which is used by insurance companies for adequately diagnosing mental disorders. Without this alignment, hospitals, counselors and other mental health practitioners cannot be reimbursed.

Other critiques of the DSM-V update include but are not limited to the distinction of alcohol abuse. In the DSM-IV the diagnostic criteria for abuse only applied to individuals who met the criteria of listed items one through four, in a 12-month period. Those who exceeded such criteria during the same time period received the dependence diagnosis. In the revised DSM-V the criteria are not as definitive. The manual states that anyone who meets two or more of the 11 listed criteria in a 12-month period can be diagnosed with alcohol use disorder. It goes on to categorize the diagnosis in three levels of severity: mild, moderate and severe. The presence of two to three symptoms, the presence of four to five symptoms and the presence of six or more symptoms constitute a mild, moderate, or severe specifier, respectively.

The DSM-V has progressed towards a vague analysis. Under the criteria it will be difficult to discriminate between first time substance abusers and recurrent addicts. The guidelines do not take into consideration the treatment needs of the clients being diagnosed. Furthermore, culture-related diagnostic issues in the DSM-V suggest that alcohol is the most frequently used intoxicating substance and contributes to considerable morbidity and mortality.

Conclusion

Another perspective that has not been discussed is the acceptance of alcohol abuse in greater society. A trip through any large city on public transportation leaves civilians vulnerable to a slew of alcohol advertisements. This is not to say that advertisements themselves have a correlation to substance abuse but one must take note of the glamorization effects. Drinking in the media is prevalent in cartoons, commercials, newscasts and various other avenues. "An unacceptably high percentage of the adult population is psychologically dependent on alcohol, according to the first in-depth study of the nation's attitudes toward drinking, conducted by the Priory Group" (Alcohol For The Soul, 2004). Drinking is almost always expected in large groups, and society appears accepting of such behavior.

Unfortunately, even as drinking is socially acceptable, alcoholism is not. Society promotes alcoholism through sexualized ads of stunning women and men, with a brief advisement to drink in moderation. "Research also reveals a picture of socially-acceptable binge-drinking, with alcohol misuse especially prevalent in the upmarket population" (Alcohol for the Soul, 2004). Sobering commercials against drinking and driving are aired by certain organizations, while some states sponsor freeway death counts of alcohol related traffic accidents. The same bars that advertise unbeatable drink deals also provide a parking lot for patronizing. The simple laws of supply and demand can explain how the accessibility of alcoholic substances has generated increasing demand.

There also appears to be a connection between the economic trends of alcohol consumption and mental health diagnosis. The DSM-V has created a general description of alcohol use disorder which may or may not create trends that could become costly as a result of misdiagnosed individuals. As the alcohol industry continues to grow in revenue and economic gains so does the prevalence of abusers. It could be suggested that the healthcare field could benefit from significant adverse effects in correlation to the industry consumption statistics. The definition of alcoholism is too general and exploits the general population for financial gain.

Over time, alcohol has grown farther away from medicinal and health improving uses. In the historic past of alcohol consumption society has typically followed cultural trends. If our society moved towards the benefits of alcohol opposed to glamorizing abuse in social settings it may decrease the self-medication facts. Self-medicating — using alcohol to fill the "hole in the soul" created by depression, fear, loneliness, trauma, low self-esteem and other mental health issues – has replaced medicinal use of alcohol (Alcohol for the soul, 2004). There is undoubtedly a love-hate relationship that America as a society tolerates. We love to hate alcoholics while celebrating and promoting alcoholism.

References

A Brief History of Alcohol & Alcoholic Beverages - Drug-Free World. (n.d.). Retrieved April 15, 2015, from http://www.drugfreeworld.org/drugfacts/alcohol/a-short-history.html

Alcohol for the soul. (2004). *CPJ: Counselling & Psychotherapy Journal*, *15*(10), 10.

American Psychiatric Association. (2013). *Diagnostic and statistical manual of mental disorders* (5th ed.). Washington, DC: Author.

Distilled Spirits Council of the United States. (n.d.). Retrieved April 16, 2015, from http://www.discus.org/economics/

Mart, S. M. (2012). Top priorities for alcohol regulators in the United States: protecting public health or the alcohol industry?. Addiction, 107(2), 259-262. doi:10.1111/j.1360-0443.2011.03682.x

Moss, H. B. (2013). The Impact of Alcohol on Society: A Brief Overview. *Social Work In Public Health*, *28*(3/4), 175-177.

Stolberg, V. B. (2006). A Review of Perspectives on Alcohol and Alcoholism in the History of American Health and Medicine. *Journal Of Ethnicity In Substance Abuse*, *5*(4), 39-106. doi:10.1300/J233v05n04-02

www.ingramcontent.com/pod-product-compliance
Lightning Source LLC
Chambersburg PA
CBHW030006190526
45157CB00014B/444